Logical Methods

Roger Antonsen

Logical Methods

The Art of Thinking Abstractly and Mathematically

 Springer

Roger Antonsen
Department of Informatics
University of Oslo
Oslo, Norway

Original Norwegian edition published by Universitetsforlaget, Oslo, Norway

ISBN 978-3-030-63776-7   ISBN 978-3-030-63777-4   (eBook)
https://doi.org/10.1007/978-3-030-63777-4

Mathematics Subject Classification (2020): 00-01, 00A06, 03-01, 05-01, 06-01, 20-01, 68-01

Cover illustration: Six Perfect In-Shuffles With 125 Cards and Five Piles. By Roger Antonsen

This Springer imprint is published by the registered company Springer Nature Switzerland AG
The registered company address is: Gewerbestrasse 11, 6330 Cham, Switzerland

Contents

Preface

Welcome to *Logical Methods*. This book is written for those of you who like to think about and understand things. It is intended as an introduction to scientific and mathematical thinking, and it is a good fit if you have begun or are considering college or university study. The book requires very little background knowledge, so as long as you are able to understand what is written and are interested and willing to work at it, you should be able to master everything that is included in this book. More specifically, you do not need any of the mathematics you learned in school beyond the basics.

School mathematics and mathematical thinking. The mathematics taught in school is often quite mechanical and rule-based. In this book, the focus is on mathematical thinking, understanding, and proving statements ourselves. This is not the same as performing calculations, manipulating symbols, or substituting values for variables in formulas. It is more about discovering patterns, reasoning things out, finding counterexamples, and thinking logically. It is about finding out for yourself what is true, and then arguing for and proving well-formulated statements. This is a way of thinking and intellectual endeavor that is extremely creative, challenging, and addictive. Thinking logically and systematically is also useful otherwise in real life, and the goal of this book is to make you better at it. What you learn here will be a theoretical foundation that you can build on, and you are guaranteed to meet many of the concepts again later.

A logical construction. This book is a logical construction; we will build almost everything from the bottom up, without many assumptions, and try to think about everything thoroughly and accurately. If you encounter a word or a term that you do not understand, my intention is that it should either have been defined or explained earlier in the text, or be something from our language that is so common and well known that it does not require a definition. It is therefore important that you ask yourself questions along the way: *Why is it like that? What does this mean? What is this a property of?* But this also means that you have to read the text carefully, because there are many dependencies. When something is mentioned, be aware that you are likely to encounter something based on it later in the text.

Understanding. The goal of this book is for you to achieve greater *understanding*, but what does it really mean to understand something? The Hungarian-American mathematician *John von Neumann* (1903–1957) is known to have said, "In mathematics you do not understand things. You just get used to them."

Here are some perspectives on understanding that may be useful to you in reading the book, as well as in your further studies.

Understanding through repetition and plenty of time. Understanding and insight come in waves, and it takes time to abstract and generalize. Something that at first glance may seem chaotic and confusing may after a while prove to be full of patterns, structures, and connections. In order to gain understanding and insight, you need to read the definitions and examples carefully, solve plenty of exercises, all on your own, and give the material plenty of time to sink in.

Understanding through details or through overview. An interesting question is whether it is best to learn something, and gain understanding, "from below" or "from above." By from below, I mean that everything is defined initially in terms of its smallest components – from basic words, symbols, and concepts – and then upward through definitions, compositions, and structures. By from above, I mean through intuition, images, and examples, where the details are filled in gradually; this is usually much less formal and more like viewing the world from a bird's perspective. Some prefer to build everything up from below, and others prefer the opposite. I have tried to find an intermediate way here.

Understanding through experimentation. *You are hereby encouraged to doubt and question absolutely everything.* I want this book to be a kind of sandbox where it is possible to explore abstract concepts and lines of reasoning with little at stake. It is possible to gain much understanding and insight through experimenting, exploring, and playing with the subject in this way, and what is learned in this way is transferable to other situations in which precise reasoning and scientific thinking are important. *You can make any assumptions, but you must then also assume their consequences.* The interesting thing is what follows from what you assume. If you would like to assume that infinite sets do not exist, or that a particular claim is true, that is fine. But you must be aware that you must also accept the consequences of such assumptions, everything that *follows* from them.

Understanding through opposites. We can approach a new phenomenon positively, by looking at properties that apply to the phenomenon, but we can also approach it negatively, by looking at the properties that do not apply. Thinking away something – thinking the opposite – may give rise to better understanding. Therefore, in practice, try to look into a phenomenon both positively and negatively: in order to understand what X is, think about what the opposite of X is. *If you do not know the extremes or boundaries of a phenomenon, do you really know the phenomenon?*

Logic. In addition to being an introduction to mathematical and scientific thinking, this is a textbook on logic. There are many reasons that it is both rewarding and useful to learn logic. Here are some of them.

Logic, syntax, and semantics. Logic gives us training in separating syntax and semantics from each other. Here, syntax refers to characters, symbols, strings, and formal languages – what it is that represents something, while semantics refers to interpretation, meaning, models, and reference – what is being represented. *We distinguish sharply between syntax and semantics.* Some of the essence of mathematics is to represent things with symbols. Without a good understanding of what is syntax and what is semantics, it becomes very difficult to reason about these things in a good and correct way. Once represented in a logical language, many possibilities and areas of application open up.

Logic, applications, and algorithms. Knowledge representation and reasoning about knowledge are becoming increasingly important, especially because of the development of semantic technologies. Although the subject of logic is very old, perhaps it is just now, with the emergence of powerful computers, that we see the full potential of logic as a field of study. Logical methods are used today both in the theoretical analysis of algorithms, databases, and programming languages and to create practical methods and tools of great utility.

Motivation. You are likely to ask yourself, *What can this be used for? Why should I learn this?* One answer is that the study of logical methods provides a flexible toolbox with techniques that can be used for many different things. But exactly because it is so abstract, it can be difficult to see the applications immediately. It is a bit like contemplating a battery or a light bulb: there is a great diversity of applications, and how each is used depends on the context. Another answer is that you are learning mathematical *thinking*. The British-American mathematician *Keith Devlin* (1947–) uses a car as a metaphor to explain this: school mathematics is like learning to drive a car, while university mathematics is like learning about how cars work, how they can be repaired, and how you can design and build your own cars. A third answer is that it is excellent training, which both sharpens our minds and enables us to reason better. By studying the topics in this book, solving the exercises, and figuring things out for yourself, you are practicing something you will do for the rest of your life: learning and understanding.

Context and scope. The book is based on the lecture notes for the course *Logical Methods for Computer Science*, which I teach at the Department of Informatics at the University of Oslo. I cover most of the content of the book during one twelve-week semester. This means about two chapters per week.

There is much that does not fit into a book like this, and there are many chapters that give only a taste of entire disciplines with their own cultures, jargon, and introductory books. My goal is not to get to the bottom of everything in each chapter, but to give you a glimpse of a fascinating and beautiful mathematical world and to give you a solid foundation for continuing reading on your own.

The structure of the book.   The book is divided into many small chapters, with exercises at the end of each, and the intention is that each chapter should constitute an appropriate amount of new material for either one lecture or one work session. The presentation of a new topic is almost always through motivation, definition, examples, and discussion. The most important here are the definitions. There are 137 of them in total, and they make up the essence of the text; make sure you understand all the definitions! The book is intentionally free of theorems, lemmas, corollaries, and excessive numbering, and the reason is simply that they are not needed. In the back of the book you will find an index and a list of symbols that appear in the book.

Ordinary words and typographical conventions.   I have tried to use ordinary words where possible and to keep the use of technical words to a minimum. The exceptions are the words that are widely used in the literature, and which are good to know, or which you are likely to encounter in future studies.

Q   I have tried to give examples, both positive and negative, for all words, concepts, and objects that are introduced. There are 204 examples in total, and the purpose of them is to increase understanding and clarify any misunderstandings. They are marked with the symbol **Q** in the margin.   ◆

❷   The book is full of exercises that you can and should solve yourself. There are 64 of them interspersed in the text, and they are marked with the symbol **❷** in the margin. At the end of each chapter there are also exercises, most often presented in order of increasing difficulty. There are 352 such exercises in total, and they are designed to allow you to practice what you have learned.   ◆

❶   Some of the exercises in the text have suggested solutions, marked with the symbol **❶** in the margin. But always try to solve an exercise yourself, all on your own, without checking the solution. *If you do not try, how will you know whether you would have managed to solve it?* If a suggested solution is not provided, the intention is that you should try yourself and that the exercise is not too difficult.   ◆

The symbols ◆, ◆, and ◆ mark the end of examples, exercises, and solutions, respectively.

> **Digression**
>
> The book contains 38 boxes with *digressions* like this. These are anecdotes, stories, and facts that are more or less relevant. Most elaborate on the topic discussed in the chapter, while others are just curiosities. Some are easy to understand, while others are quite demanding. Common to them is that nothing in the text depends on them, and it is therefore safe – and maybe smart, if you are reading the text for the first time – to skip them. They are mainly meant to add spice, inspiration, and recreation!

Suggested solutions. *Are there no suggested solutions for the exercises at the end of the chapters?* No. Think of the exercises as training. Reading a solution without attempting to solve an exercise yourself is like having someone perform your daily workout for you. You will not get stronger in this way. If you can't solve an exercise, do not give up. Please try again. And again. And one more time.

Dependencies between the chapters. The following is a rough overview of how the chapters depend on each other; for example, you may read Chapter 6 without reading Chapters 2–5 first. Chapters 0, 1, and 6–12 constitute a mathematical foundation with the most basic concepts and definitions you need to do mathematics on your own, and you can look at these as a mini-course in mathematical methods. Chapters 2–5, 13–16, and 24 are about logic; they deal with basic logical concepts and methods. Chapters 17–23 deal with different mathematical topics, such as combinatorics, graph theory, algebra, and language theory. Chapters 2–5 are about propositional logic and proof methods; they form a foundation for the rest of the book. My recommendation is to read the chapters in order until you come to Chapter 7. From there, you have much more freedom in choosing your path.

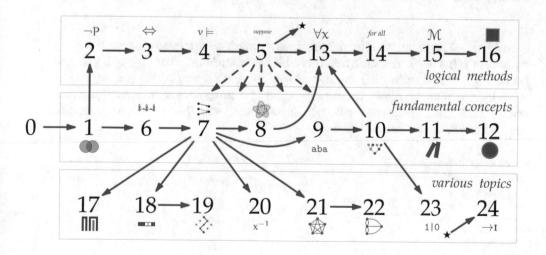

Thank you. This book is the product of years of teaching, and I would like to thank everyone who has provided constructive feedback, comments, exercises, solutions, and lots of inspiration along the way. This book could not have existed without you. I would like to thank Springer for supporting me and being patient with my translation efforts, and Universitetsforlaget in Norway for supporting the first version of the book. Much of the contents of this book have been created in dialogue with creative students, sharp teaching assistants, and skilled co-lecturers over many years: Thank you to all of you; it is a privilege to work with you. I would also like to thank colleagues, friends, family, and everyone who has supported me. You are a wonderful mix of family, students, teaching assistants, colleagues, logicians, and friends. A special thanks to: *Kelly Nelson, Karin og Ingvald Antonsen, Stål Aanderaa, Helmer Aslaksen, Julia Batkiewicz, Peter Brottveit Bock, Jens Erik Fenstad, Jon Henrik Forssell, Martin Giese, Håkon Robbestad Gylterud, Christian Mahesh Hansen, Sigmund Hansen, Knut Hegna, Herman Ruge Jervell, Einar Broch Johnsen, Sofie Magdalena Jønnvoll, Leif Harald Karlsen, Johan Wilhelm Klüwer, David Kramer, Lars Kristiansen, Dag Langmyhr, Espen Hallenstvedt Lian, Rémi Lodh, Håkon Salomonsen Møller, Andreas Nakkerud, Dag Normann, Olaf Owe, Jonathan Ringstad, Ragnhild Kobro Runde, Martin Georg Skjæveland, Martin Steffen, Lars Kristian Maron Telle, Trond Thorbjørnsen, Evgenij Thorstensen, James David Trotter, Lars Tveito, Eli Valheim, Arild Waaler*, each of whom in their own way has contributed to the book you now have in front of you.

Good luck. I hope you will learn lots of mathematics, logic, and abstract thinking by reading this book. Imagine that it is a bit like mountain climbing; you will not become good at it without *doing* it, and it is equally rewarding *for you* no matter what level you are at. Play with the subject and be curious; assume something and see where it takes you. *Perhaps you will discover something that no one else before you has discovered. Good reading/thinking!*

■■■■■■■■■■■■■■■■■■■■■■

Oslo, August 2020
Roger Antonsen

Chapter 0

The Art of Thinking Abstractly and Mathematically

In this chapter we look at some basic concepts that we will meet over and over again: truths, definitions, assumptions, proofs, languages, axioms, and theorems. We also lay some groundwork and make some conceptual clarifications for the road ahead.

Abstraction

What does it mean to think abstractly? Thinking abstractly, systematically, and mathematically is something we all do every day. You could not, for example, read or calculate if you did not have the ability to abstract. You recognize combinations of words and numbers because you have abstracted over how they are used. On the other hand, abstraction is something you can become better at and reason about. The aim of this book is to make you better at abstraction, as well as thinking abstractly, systematically, mathematically, and scientifically. *Do you see what the next number is?*

$$1 \quad 3 \quad 4 \quad 7 \quad 11 \quad 18 \quad 29 \quad ?$$

Numbers, however, are only one type of mathematical object; we also have *sets, formulas, terms, expressions, relations, functions, strings, graphs, trees, calculi,* and many others. And we can work with and reason about these mathematical objects in the same way we do with numbers; we can define them and study how they are combined, compared, and categorized. This is part of the abstract nature of mathematics: we apply mathematical reasoning to things that are much more general than numbers. *For example, do you see what the next figure is?*

     ?

To abstract is also to think away what is inessential, what is irrelevant or not of interest in a given context; it is the practice of focusing on particular aspects of something and ignoring irrelevant details.

© Springer Nature Switzerland AG 2021
R. Antonsen, *Logical Methods*, https://doi.org/10.1007/978-3-030-63777-4_1

Reasoning About Truth

Why is 2 + 3 = 3 + 2 true? Do infinite sets exist? How do we know that something is true? What does it mean that something is true? Is it true at all times and in any place? How can we prove that something is true?

These questions are fundamental in science and mathematics. Here, we are going to reason logically and mathematically, not just read about what is true. We are going to practice discovering and reasoning about truth, even *proving* that something is true. It's the difference between an *observer* and a *participant*. When you prove something, you often feel that something "clicks" and "falls into place," which can be both rewarding and addictive.

A proof makes us absolutely sure that something is true, and it gives us the ability to convey our discoveries and insights to others. When we have proved something, we know that it is true, not just here and now, but forever and anywhere in the whole universe. But what is true depends on what we assume.

Assumptions

In mathematical reasoning, we begin with *assumptions*. We are interested in what follows, what is true, and what can be proved from a set of assumptions. Logical arguments are used to draw conclusions – often called **theorems** – from basic assumptions – often called **axioms**. What we assume is therefore crucial. If we are sure of our assumptions, we also want to be sure of what we derive from those assumptions. That is why the study of what *logically follows* from a set of assumptions is so important. If we assume that something is true, we want a guarantee that everything that follows is also true. In this book, we are going to practice *thinking from assumptions*. There are many historical examples of theories about whose assumptions one was completely sure, until it was discovered that the assumptions could be changed to obtain different, equally valid, theories. The parallel axiom in Euclidean geometry is one such example.

Digression

Euclid was a Greek mathematician who lived about 2300 years ago. He is known for what we call *Euclid's Elements*, a collection of 13 books on mathematics and geometry. This is history's first example of the use of the axiomatic method. In the first of these books, 10 axioms are presented, and the fifth of these is the *parallel axiom*, also called the *parallel postulate*. This essentially states that if a straight line and a point external to the line are given, there is exactly one straight line in the plane determined by the point and the line that passes through the point and does not intersect the given line. Changing this axiom so that it says something other than "exactly one straight line" gave rise to so-called *non-Euclidean* geometries. By saying "no straight lines," we got one type of geometry, *projective* geometry; by saying "infinitely many straight lines," we got another, *hyperbolic* geometry.

Language

All mathematical reasoning is done in a mathematical *language*. How formal and symbolic this language is ranges from colloquial oral and written language to machine-readable source code. We can use this entire spectrum in communicating our reasoning. Having a common language and an understanding of what words and concepts mean enables us to communicate effectively with each other. We wish to convey thoughts, reasoning, insights, and proofs without room for misinterpretation. In order to do that, it is necessary to have a clear and precise language without ambiguities. Good definitions give us that.

Definitions

A **definition** tells us how words, symbols, or terms are to be understood and used. While you may think that definitions are just something that we look up in a dictionary, in many theoretical disciplines, such as mathematics and computer science, definitions play a special and important role.

A definition determines and provides limits for what it defines. If a word or term is not defined, we cannot really use it in a meaningful way. When a word is defined, we have a starting point for a common understanding. We are going to write definitions in the following way:

> **Definition 0.1. Googol**
>
> A **googol** is the number 10^{100}, a one followed by a hundred zeros. A **googolplex** is the number 10^{googol}, a one followed by a googol zeros.

Both assumptions and definitions have a certain arbitrariness about them, in the sense that they could have been different. It is important to be aware of this arbitrariness. But when the assumptions and definitions are finally given, it is they that rule. Once a word is defined, we can no longer discuss what the word means – that is given by the definition – but we can always discuss whether the definition is a good one. It is also important to note that a definition is always applied in a particular context; for example, two scientific articles may define one and the same word completely differently. However, there are many words, symbols, and concepts that are defined in the same way, and it is useful to get to know many of these.

A definition is something that separates and sets a boundary between what falls within a concept and what does not. For this reason, it is often instructive to ask what is *not* covered by a definition. When we read a definition of a property, it pays to find many examples of things that have the property and many examples of things that do *not* have the property. We then approach the phenomenon from both sides, both positively and negatively.

Proofs

Once we have our language, our definitions, and our assumptions in place, we can find proofs. A **proof** of a statement is a line of reasoning that convinces us that the statement is true given certain assumptions. This usually consists of a number of logical inferences that show how to get from the assumptions to the statement.

For each step in the line of reasoning, the conclusion must be a *logical consequence* of the assumptions. That something is a logical consequence means that it follows, with logical necessity, from what we assume. When we have proof of a statement, we know that the statement is at least as true as the assumptions. This requires, however, that all the steps in the proof be correct. A proof often contains a large number of inferences, and it is enough for one step to be incorrect for the whole proof to be incorrect. We must therefore be accurate, and constantly check that each new conclusion is a logical consequence of what has previously been established. Here, we will learn to write proofs correctly and understand how a line of reasoning can be flawed, and in that case, why it is flawed. This is also the beginning of a study that is both mathematical and philosophical in nature. A big and difficult question, which is currently unsolved, is the question of when two proofs of the same statement can be said to be the same. *How can proofs be equated?*

Problem Solving and Pólya's Heuristics

In mathematics and computer science, we solve problems all the time, from simple calculations to complex and compound problems. It is therefore a good idea to reflect a bit on this process. *What do we do when we get stuck and are unable to solve a problem?* This question is useful to think about in most subjects, and elsewhere in life, and you might find the following heuristics, or *rules of thumb*, useful. They are taken from a classic book from 1945 called *How to Solve It?*, written by the Hungarian mathematician *George Pólya* (1887–1985). He divides the process into four phases (although they often overlap):

(1) First, you need to *understand* the problem.
(2) Attempt to use previous experience from similar problems to make a *plan*.
(3) Execute the plan.
(4) Look over and check whether you actually believe in the answer you got.

Pólya's heuristics are not meant to be a recipe that you can follow exactly and that always give you an answer, nor will they necessarily make you a master of problem solving, but they are practical tips that may be of great help if you are stuck on a problem that you are trying to solve. The heuristics are formulated as a series of questions that we can ask ourselves or others, and each question is designed to help us get closer to a solution.

(1) Understand the Problem

First, you need to *understand* the problem. Do you understand all the words in the statement of the problem? Perhaps you have to check the definitions of the words used. *What is the unknown? What is given? What are the assumptions?* Are the assumptions satisfiable? Are the assumptions sufficient to find the unknown? or insufficient? or redundant? Draw a *figure* or illustration, and find a suitable notation. Divide the assumptions into smaller pieces. Can you express the problem in your own words? This phase may be the most important, because when you really understand what the problem is, you are already a bit closer to a solution.

(2) Make a Plan

This is about finding the connection between the assumptions and the unknown, and you should come up with a plan to find a solution to the problem. *Have you seen the problem before?* Or have you seen the same problem just in another form? *Have you seen a similar problem?* Do you know a result that might be useful? *Look at the unknown!* Try to remember a problem with the same unknown. *What is the connection between the assumptions and the unknown?*

This should lead to a *plan* to find a solution. Some strategies for making a plan could be the following: *Draw a figure. Eliminate possibilites. Look at special cases. Look for a pattern.* Think backwards: What is needed to get to the unknown? Be creative and think differently.

If you cannot solve the problem: *Can you solve a simpler problem?* Can you change the assumptions such that you can solve it? Can you solve a more general problem? Or a more specific problem? Or can you solve part of the problem? If you drop any of the assumptions, how close do you still get to the unknown? What happens if you change the assumptions or the unknown? Can you deduce something useful from the assumptions? Can you come up with other assumptions that would have helped? Did you use all the assumptions?

(3) Execute the Plan

Execute the plan and check each step! Is each step correct? Can you prove that each step is correct? If your plan didn't work out, make another plan.

(4) Look Over and Check

Check that the result is correct. Check that the reasoning holds. Do you believe it yourself? Is it understandable? Can you answer the problem in another way? Is the solution intuitively convincing? Do you see the solution at once? Does the solution agree with the assumptions? Can you use the result, or the method you used, on another problem?

Digression

Here are some puzzles for you to ponder.

(1) You have two fuses. Each fuse burns up in exactly one minute, but they burn unevenly and unsteadily. *Can you measure a time interval of exactly* 45 *seconds using these two fuses? Can you measure* 10 *seconds?*

(2) You have a bag of red and green balls. After picking out four balls completely at random, and each was green, you are informed that the likelihood of this happening was exactly 50%. *How many red and green balls are left in the bag?*

(3) Along a large circular race track, a certain number of gas stations are located at random intervals. The amount of gas needed to drive exactly one lap around the track is distributed between the gas stations. *Show that there is a starting point that lets you drive around the entire track*, given that you start with an empty tank and that the tank can accommodate all the gas that is distributed around the track.

(4) You have two hourglasses, one that can measure three minutes and one that can measure five minutes. *How can you use these two hourglasses in cooking an egg for four minutes?*

(5) Suppose we have made a 3×3×3 cube out of 27 smaller cubes and that an ant is sitting on the outside of the cube, in the center of one of the faces of a small cube. It bores its way to the center of that cube, and then bores from the center of a cube to the center of an adjacent cube. *Is it possible for the ant to eat its way through all the outer cubes and ultimately end up in the central cube without ever visiting the same cube twice?*

(6) You are placed in a 50 m × 50 m × 50 m box that stands on four 100-meter-high columns. From the middle of the ceiling inside the box hang two ropes, both of which are exactly 50 meters long, and they are attached to the ceiling with exactly one meter between then. The box has only one opening, a small window at the bottom of one wall, where there is a small hook as well. You have a knife available, but no clothes or other aids. The farthest fall you are able to survive is 10 meters. *How can you manage to get down to the ground alive?*

(7) One hundred prisoners are sitting in a bus that will take them to a prison where they will spend the rest of their days in isolation, without the opportunity talk to anyone other than the prison guards. In this prison there is a room with two switches, and the prisoners will be sent into this room in an order, one by one, decided by the guards. All the prisoners will sooner or later be sent into the room, even if they have been in there several times. When the prisoners are in the room, they must flip exactly one of the switches, and each switch has exactly two states, on or off. The prisoners are released if one of them with certainty can state that everyone has been in the room, but they will all be executed if someone states it incorrectly. *What strategy should the prisoners use to ensure their freedom?*

(8) *What is the next number?*

$$100 \quad 121 \quad 144 \quad 202 \quad 244 \quad 400 \quad ?$$

Chapter 1
Basic Set Theory

▮▪▪▪▪▪▪▪▪▪▪▪▪▪▪▪▪▪▪▪▪▪▪▪

In this chapter you will learn the basic concepts and notation used in set theory: what a set is; what operations we perform on sets, such as intersection, union, and set difference; and how sets can be constructed and compared. You will also learn about tuples and Cartesian products.

First Steps

We have to start somewhere, and there are many reasons why we start with set theory. First of all, we use concepts from set theory all the time in many theoretical subjects as well as practical applications, and it is a good idea to familiarize yourself with many of these concepts. Basic set theory also constitutes a minimum theoretical vocabulary for speaking precisely about many mathematical concepts and constructions. It also helps us to practice thinking abstractly and mathematically about things other than numbers. A deeper and more important reason why set theory is important is that set theory is a foundation for all of mathematics. You can define most things in mathematics based on the concept of sets.

What Is a Set?

Let us now take a look at our first definition. Once you understand it, you will know what a set is.

Definition 1.1. Set

A **set** is a finite or infinite collection of objects, where we ignore the order and number of occurrences of each object. The objects in a set are called **elements**. If x is an element of the set A, we write $x \in A$. If a is *not* an element of A, we write $a \notin A$. Two sets A and B are **equal** if they contain exactly the same elements, and then we write $A = B$. We write $A \neq B$ if they are not equal. A set can be specified by writing the elements between the symbols $\{\ \}$, which are often called curly braces.

One of the first things we see is that the word "set" is defined using the word "collection" and that this latter word is not defined further; we take it for granted

© Springer Nature Switzerland AG 2021
R. Antonsen, *Logical Methods*, https://doi.org/10.1007/978-3-030-63777-4_2

that we understand what it means. Similarly, the definition makes use of other undefined terms, such as *order*, *object*, *finite*, *infinite*, and *occurrence*. In any definition, there is always something we must take for granted, what is important is that nothing should be unclear. We may use examples to illustrate what we mean. If we go further into the mathematical discipline of set theory, we can define sets more precisely, but for the time being, we will assume that our understanding of language is sufficient to understand the concept.

Q   Sets. Let us take look at some simple sets. We have $\{0, 1\}$, which is the set of values that a **bit** can have. The set $\{o, 1, i, g, k, s\}$ consists of six characters, and the set $\{5, 6, 7, p, q, r\}$ consists of both numbers and characters. The set $\{2, 3, 5, 7, 11, 13, 17, 19, 23, 29\}$ consists of the 10 smallest prime numbers. The set of all people under the age of 18 is also a set.   ♦

Q   Number sets. Here are some well-known and much-used number sets:

- $\mathbb{N}$ is the set of **natural numbers**: $\{0, 1, 2, 3, \dots\}$.
- $\mathbb{Z}$ is the set of **integers**: $\{\dots, -3, -2, -1, 0, 1, 2, 3, \dots\}$.
- $\mathbb{Q}$ is the set of **rational numbers**, or **fractions**, which are numbers that can be written in the form $\frac{m}{n}$, where m and n are integers and n is not equal to 0.
- $\mathbb{R}$ is the set of **real numbers**, which represents the points on a continuous number line. This set contains, for example, such irrational numbers as e, π, and $\sqrt{2}$.   ♦

Q   Sets and elements. The definitions of "set" and "element" tell us how we should use the symbols $\in$ and $=$. We have that $1 \in \{1, 3\}$, because 1 is an element of $\{1, 3\}$, but that $2 \notin \{1, 3\}$, because 2 is *not* an element of $\{1, 3\}$. We have that $\{1\} \neq \{1, 2\}$, because the second set contains an element that is not in the first. It is also the case that the order of the elements does not matter; the sets $\{1, 2\}$ and $\{2, 1\}$ are identical. Neither does the number of occurrences make any difference; the sets $\{1, 2, 2\}$ and $\{1, 2\}$ are also identical, even though the number 2 is written twice in $\{1, 2, 2\}$.   ♦

Here, we see our first example that there is a distinction between a representation and what is being represented. In the example, there are three different representations, $\{1, 2\}$, $\{2, 1\}$, and $\{1, 2, 2\}$, of the same underlying set, namely the set that consists of the numbers 1 and 2. When the definition says $a \in A$, it means that A *represents* a set and that a *represents* something that is an element of this set.

Q   Sets. The following sets are all identical and consist only of 1, 2, and 3:

$$\{1, 2, 3\} = \{3, 1, 2\} = \{3, 3, 1, 2\} = \{3, 3, 1, 1, 2\} = \{3, 2, 3, 1\}.   ♦$$

We can also have empty sets and sets of sets. In axiomatic set theory, we usually start with the empty set and build up all sets from there.

Definition 1.2. The empty set

The **empty set** is the set that contains no elements. The empty set is written $\{\}$ or $\varnothing$.

The descriptions "the set of odd numbers divisible by 2" and "the set of people who are one hundred feet tall" both refer to the empty set, because there is nothing that has these properties.

❷ Explain why $\varnothing$ and $\{\varnothing\}$ are different.   ◆

❶ The first set, $\varnothing$, is empty; it has no elements. We can also write $\{\}$ for this set. The second set is *not* empty; it has one element, namely $\varnothing$. We can also write $\{\{\}\}$ for this set. Similarly, a bottle containing an empty bottle empty is not empty.   ◆

> **Digression**
>
> The symbol for the empty set, $\varnothing$, originates from the Norwegian letter Ø. It was the French mathematician *André Weil* (1906–1998), known as one of the people behind the pseudonym *Bourbaki*, who introduced this symbol in 1939 because he knew about the Norwegian alphabet. It was also the Bourbaki group who introduced the words "injective," "surjective," and "bijective," which we will learn about in the chapter on functions.

Building Sets

One way to specify a set is by writing the elements between $\{$ and $\}$, but that is often cumbersome and impractical. The following is a more practical way to do it.

Definition 1.3. Set builder

A set can be defined as the set of all elements that have a given property. Such a construction is called a **set builder** (or set comprehension or set abstraction). A definition of the form *"the set of all elements x such that x has the property P"* is written $\{x \mid x$ has the property P$\}$. What is on the left-hand side of the line can also be a complex expression.

Q **Set builder.** Here are some ways to use the set builder:

  – $\{n \mid n \in \mathbb{N}$ and $n < 108\}$ is the set of all natural numbers less than 108.
  – $\{X \mid X$ is a person over the age of 18$\}$ is the set of people who are more than 18 years old.

— $\{x \mid x \in A \text{ or } x \in B\}$ is the set of everything that is an element of A or B (or both).
♦

For our purposes, we can be quite free when it comes to what we allow to the right of the line, but we would quickly run into trouble if we allowed the property P to be just anything. One problematic example is $\{x \mid x \text{ is a set}\}$, namely the "set" of all sets, or, even worse, $\{x \mid x \text{ is } not \text{ a set}\}$, "the set" of everything that is not a set.

> ### Digression
>
> In Seville, there is a barber who shaves all of the inhabitants, and only those, who do not shave themselves. Now, the question is this: *does the barber shave himself?* After a little reflection, we see that we have an impossible situation: if he does not shave himself, then he shaves himself, and vice versa. Notice the words "and only those" in the first sentence. If the sentence had said only that he "shaves all of the inhabitants who do not shave themselves," it would not be an impossible situation because then the barber could, for example, shave everyone, regardless of whether they shave themselves. This is a variant of **Russell's paradox**, named after the British mathematician and philosopher *Bertrand Russell* (1872–1970): Let X be the set of exactly those sets that do not contain themselves as elements. Now we can ask whether X is an element of X. After a little reflection, we have an impossible situation: if X is an element of X, then X is not an element of X, and vice versa. *How do we resolve such paradoxes?* The most common way is simply to ban barbers and sets of this type.

If what is to the left of the line is a composite expression, it can be viewed as a shorthand notation.

Q **Set builder, shorthand notation.** It is common to write

$$\{x \in A \mid x \text{ has the property } P\}$$

for the set of all elements of A that have the property P. We could just as well have written $\{x \mid x \in A \text{ and } x \text{ has the property } P\}$, which means exactly the same thing. For example, the set of odd numbers can be denoted by $\{2x + 1 \mid x \in \mathbb{N}\}$, the set of elements $2x + 1$ such that x is a natural number; in other words, for every natural number x, the number $2x + 1$ is in the set (and nothing else). This means exactly the same as $\{y \mid y = 2x + 1 \text{ and } x \in \mathbb{N}\}$, but it is somewhat easier to read.   ♦

It is also possible to construct sets that contain other sets.

Q **Sets of sets.** The following shows some ways in which sets may consist of other sets. Make sure you understand each point before you continue.

— $a \in \{a\}$, but $a \notin \{\{a\}\}$. The object a is an element of $\{a\}$, but it is *not* in $\{\{a\}\}$; the only element of the set $\{\{a\}\}$ is the set $\{a\}$. We cannot "see into" sets that are elements of other sets.

- $\{\{p \mid p \text{ is a potato}\}\} \neq \{p \mid p \text{ is a potato}\}$. A sack containing a sack of potatoes is not in itself a sack of potatoes, but only a sack containing one element, namely a potato sack.
- $\{\{0\}, \{1\}, \{2\}, \dots\} = \{\{n\} \mid n \in \mathbb{N}\}$. This is a set of sets; each element is a set that consists of one natural number. Notice the difference between this set and $\{n \mid n \in \mathbb{N}\}$, which is just $\mathbb{N}$.
- $\{\{0, 2, 4, 6, \dots\}, \{1, 3, 5, 7, \dots\}\}$ is a set of two elements: one element is a set of even numbers and the other element is a set of odd numbers. ◆

Notation. If $a, b, c, \dots$ are elements of A, we may write $a, b, c, \dots \in A$ instead of $a \in A, b \in A, c \in A$, etc.

Operations on Sets

Now that we have defined sets, we are going to look at how we can create new sets from old sets, then how we can visualize sets and compare them to each other. We will see that this is something we do for many different types of structures, not just sets: we define the objects, create operations on the objects, and look at how the objects can be compared and categorized. This is, for example, exactly what we do with numbers: we define numbers, use the ordinary arithmetic operations to create new numbers, and compare them using, for example, the equality relation and the less-than relation. In Chapter 7, we will define precisely what is covered by the concept of an *operation*.

Definition 1.4. Union

The **union** of two sets A and B is the set that contains exactly those elements that are elements of A or B, including the elements that are contained in both. This means that all elements of A and all elements of B, but no other elements, are elements of the union. The union of A and B is written $(A \cup B)$. We allow ourselves to drop the parentheses and write $A \cup B$, as long as nothing becomes ambiguous.

Q **Union.** The union of $\{a, b\}$ and $\{c, d\}$ is $\{a, b, c, d\}$, and the union of $\{a, b\}$ and $\{b, c\}$ is $\{a, b, c\}$; the element b occurs both in $\{a, b\}$ and $\{b, c\}$, but is counted only once. The union of a set and the empty set is always the set itself: $\{1, 2, 3\} \cup \varnothing = \{1, 2, 3\}$. ◆

Definition 1.5. Intersection

If A and B are sets, the **intersection** of A and B, or A **intersected with** B, is the set containing exactly those elements that are elements of both A *and* B. The intersection of A and B is written $(A \cap B)$. We allow ourselves to drop the parentheses and write $A \cap B$, as long as nothing becomes ambiguous.

Q **Intersection.** The intersection of $\{a, b\}$ and $\{b, c\}$ is $\{b\}$, because b is the only element the sets have in common. If two sets do not have any element in common, their intersection will be empty: $\{a, b\} \cap \{c, d\} = \varnothing$. When we take the intersection of something and the empty set, we get the empty set: $\{1, 2, 3\} \cap \varnothing = \varnothing$, because there is nothing that is in both $\{1, 2, 3\}$ and the empty set.   ♦

Definition 1.6. Set difference

If A and B are sets, then **the set difference** of A and B, or A **minus** B, is the set that contains exactly those elements that are elements of A but *not* elements of B. Equivalently, it is the set that contains all elements remaining in A after all elements of B that are also in A have been removed. The set difference of A and B is written $(A \setminus B)$. We allow ourselves to drop the parentheses and write $A \setminus B$, as long as nothing becomes ambiguous.

Q **Set difference.** We have $\{a, b\} \setminus \{a\} = \{b\}$; here, the element a is removed from $\{a, b\}$. On the other hand, if we take $\{a, b\} \setminus \{c, d\}$, we get $\{a, b\}$. Because c and d are the only candidates for removal, and neither of them is in $\{a, b\}$, the result is still $\{a, b\}$. When we take the set difference of something and the empty set, it has no effect: $\{1, 2, 3\} \setminus \varnothing = \{1, 2, 3\}$; if we have something and do not remove anything, we will be left with what we had before. And no matter what elements we might think of removing from the empty set, they will not be in the empty set, and so we will still have the empty set: $\varnothing \setminus \{a, b, c\} = \varnothing$.   ♦

Visualizing Sets

There are many ways to visualize sets. **Venn diagrams**, named after the British logician and philosopher *John Venn* (1834–1923), is one such way. A Venn diagram shows the different ways that sets can be combined using the operations we have learned: union, intersection, and difference. Each set is represented as a closed region, and what is inside the region represents the elements of the set.

If we first look at two sets A and B, we can represent A with the figure ◔ and B with the figure ◑. If we use pen and paper, we can draw a hatch pattern in the different regions. Then we can represent operations on the sets as follows. Here, the colors are only a guide; what is important is whether a region is colored or not.

The reason that Venn diagrams work so well is that every possible combination is represented as exactly one region in the figure. When we have two sets, we have four possibilities – an element is contained in both sets, in only one of the sets, or in neither of them – and thus we get four regions. We can also create Venn diagrams for three elements, and then we get eight different regions and many more combinations. Using three sets and only union and intersection, we get the following:

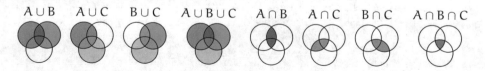

When we combine set operations, we get many more sets, and all of them can be illustrated with Venn diagrams. Here are some of them. *Can you figure out how many combinations there are altogether?*

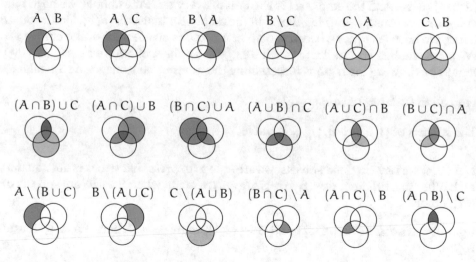

❷   Use Venn diagrams to show that $A \cup (A \cap B) = A$.   ◆

❶   We calculate the Venn diagram for $A \cup (A \cap B)$ as follows. First, we calculate $A \cap B$: ⬤∩⬤ = ⬤. Then we calculate $A \cup (A \cap B)$: ⬤∪⬤ = ⬤. We see that this set is identical to A, ⬤. This is exactly what is expressed by the following equation:

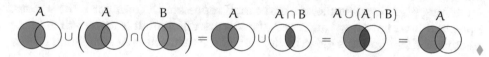

Comparing Sets

> **Definition 1.7. Subset**
>
> A set A is a **subset** of a set B, or **included** in B, if all the elements of A are also elements of B. We write $A \subseteq B$ when A is a subset of B, and otherwise, we write $A \nsubseteq B$.

Q **Subset.** We have that $\{a, b\} \subseteq \{a, b, c\}$, because all the elements of $\{a, b\}$, namely both a and b, are elements of $\{a, b, c\}$. But $\{a, b, c\} \nsubseteq \{a, b\}$, because c is an element of the first set but not the second one. Every set is a subset of itself; for example, $\{a, b\} \subseteq \{a, b\}$. And the empty set is a subset of every set; for example, we have that $\varnothing \subseteq \{a, b\}$, because everything that is in the empty set is also in $\{a, b\}$. On the other hand, we have $\{a, b\} \nsubseteq \varnothing$, because neither a nor b is an element of the empty set. We can visualize and indicate that a set ⬤ is a subset of another set ◯ as ◉, that is, by drawing the regions representing the different sets one inside the other.◆

❓ Find all the subsets of the set $\{1, 2\}$. ◆

❗ The subsets of $\{1, 2\}$ are $\varnothing, \{1\}, \{2\}$, and $\{1, 2\}$. ◆

Now that we have defined subsets, we are able to express that two sets are equal to each other in a different way: two sets are *equal* exactly when the sets are subsets of each other.

Q **Equality of sets.** If $A = \{1, 2\}$ and $B = \{2, 1\}$, then $A = B$, because $A \subseteq B$ and $B \subseteq A$. ◆

It is common to use this fact as a proof method to show that two sets are equal; if each set is included in the other, they must be equal.

Q **Equality of sets.** Are $\{a, b\}$ and $\{b, c\}$ equal? It depends! Thus far, we have been using letters as elements of sets. But it is also possible for letters to represent something else. For example, if a and c each represent the number 1, the sets are equal. But if we have $a = 1$ and $c = 2$, the sets are different. It is important to know whether the symbols in a given context stand for themselves or something else. ◆

❓ Are the sets $\{2 + 2\}$ and $\{4\}$ equal? ◆

❗ It depends on how we interpret what is between { and }. If we look only at the characters, then "2 + 2" and "4" are different, and the sets are not equal. But if we look at $2 + 2$ and 4 as *numbers*, then both are equal to the number 4, and the sets are therefore equal. ◆

Tuples and Products

Now let us look at some ways to make new sets from old ones. One way is to "multiply" sets. *What does it mean to multiply sets?* In order to define that, we must first define tuples.

Definition 1.8. Tuples

A **tuple** of n elements, an n-tuple, is a collection of n objects in which both the order and the number of occurrences of each object matter. A 2-tuple of two elements x and y is called an **ordered pair**, or simply a **pair**, and is written $\langle x, y \rangle$. A 0-tuple, the empty tuple, is written $\langle \rangle$. A 1-tuple of one element x is identified with x. Two n-tuples $\langle a_1, \ldots, a_n \rangle$ and $\langle b_1, \ldots, b_n \rangle$ are **equal** if their corresponding components are equal. In other words, $a_i = b_i$ for $i \in \{1, 2, \ldots, n\}$.

Q **Tuples.** The set $\{ \langle 1, 2 \rangle, \langle 2, 4 \rangle, \langle 3, 6 \rangle, \langle 4, 8 \rangle, \ldots \}$ is a set of tuples. We note that $\langle 1, 2 \rangle$ is not equal to $\langle 2, 1 \rangle$, because the order of the elements is different. It is not enough that the tuples contain the same elements. Similarly, $\langle 1, 2 \rangle$ and $\langle 1, 2, 2 \rangle$ are different, because they do not contain the same numbers of each element. It is also the case that $\langle 1, 2 \rangle \neq \{1, 2\}$, because a pair is not the same as a set.   ♦

Definition 1.9. Cartesian product

The **Cartesian product**, sometimes called the **cross product**, of n sets $X_1, X_2, \ldots, X_n$ is written $X_1 \times \cdots \times X_n$ and defined to be the set of all n-tuples

$$\{ \langle x_1, \ldots, x_n \rangle \mid x_i \in X_i \text{ for } i = 1, \ldots, n \},$$

in which each element x_i comes from the set X_i. The expression X^n is an abbreviation for $\underbrace{X \times X \times \cdots \times X}_{n \text{ times}}$. We let X^0 denote the set containing only the empty tuple, $\{ \langle \rangle \}$.

If A and B are sets, the Cartesian product of A and B is the set of all pairs $\langle x, y \rangle$ such that $x \in A$ and $y \in B$.

Q **Cartesian product.** The Cartesian product of $\{a\}$ and $\{1, 2\}$ is $\{ \langle a, 1 \rangle, \langle a, 2 \rangle \}$. This is not the same as the Cartesian product of $\{1, 2\}$ and $\{a\}$, which is $\{ \langle 1, a \rangle, \langle 2, a \rangle \}$. We have that $\{\bullet, \circ\} \times \{\bullet, \circ\}$ is equal to $\{ \langle \bullet, \bullet \rangle, \langle \bullet, \circ \rangle, \langle \circ, \bullet \rangle, \langle \circ, \circ \rangle \}$, and this can be written $\{\bullet, \circ\}^2$. A chessboard can be viewed as the Cartesian product of the set $\{a, b, c, d, e, f, g, h\}$ and $\{1, 2, 3, 4, 5, 6, 7, 8\}$. For each of the 64 squares, we have one letter and one number. If we take the product of X and the empty set, we get the empty set, no matter what X is; $X \times \varnothing = \varnothing$. Below are two different ways to visualize the Cartesian product $\{a, b, c, d, e\} \times \{1, 2, 3\}$. Notice that the number of squares on the "chessboard" is the same as the number of lines between the circles.

3	$\langle a,3\rangle$	$\langle b,3\rangle$	$\langle c,3\rangle$	$\langle d,3\rangle$	$\langle e,3\rangle$
2	$\langle a,2\rangle$	$\langle b,2\rangle$	$\langle c,2\rangle$	$\langle d,2\rangle$	$\langle e,2\rangle$
1	$\langle a,1\rangle$	$\langle b,1\rangle$	$\langle c,1\rangle$	$\langle d,1\rangle$	$\langle e,1\rangle$
	a	b	c	d	e

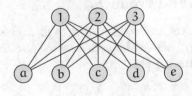

Multisets

In many contexts, we are concerned with the number of occurrences of each object, but not the order. This is commonly captured with the term *multiset*:

Definition 1.10. Multiset

A **multiset**, also called a **bag**, is a collection of objects in which the order is ignored, but not the number of occurrences of each element. A multiset can be specified by writing the elements between the symbols [ ].

A multiset is thus a "set" in which each element can occur multiple times. For example, we can write $[2,3,3,4]$ for a multiset with two instances of the number 3. In the same way as for sets, we can define terms such as union, intersection, set difference, equality, and subset for multisets. It is worth noting that there are several ways to define the union of multisets. We will not be using multisets very much as we move forward, but here are nevertheless some examples of how this can be done:

$$[1,2,3] = [3,2,1]$$
$$[1,1,2,3] \neq [1,2,3]$$
$$[1,2] \cup [1,2,3] = [1,2,3]$$
$$[1,2] + [1,2,3] = [1,1,2,2,3]$$

$$[1,1,1,2,3] \cap [1,1,4] = [1,1]$$
$$[1,1,1,2,3] \setminus [1,1,d] = [1,2,3]$$
$$[1,1] \subseteq [1,1,2,3]$$
$$[1,1,1] \not\subseteq [1,1,2,3]$$

Exercises

1.1 Let $A = \{1,2,3\}$ and $B = \{a,b\}$. Find the Cartesian products $A \times B$ and $B \times A$.

1.2 Let $A = \{1,2,3,4,5,6\}$, $B = \{1,3,5,7\}$, and $C = \{2,4,6,8\}$. Find the following sets:

    (a) $A \cup B$ (b) $A \cap B$ (c) $A \setminus B$ (d) $(A \setminus B) \setminus C$ (e) $(B \cup C) \setminus A$

1.3 Find all the subsets of $\varnothing$, $\{\bullet\}$, and $\{\bullet, \circ\}$, respectively.

1.4 Assume that $A = \{1, 2, \{1,2\}, \{1,3\}, \{1,2,3\}\}$ and determine whether each of the following statements is true:

    (a) $1 \in A$
    (b) $2 \in A$
    (c) $3 \in A$
    (d) $\varnothing \in A$

    (e) $\{1\} \in A$
    (f) $\{1,3\} \in A$
    (g) $\{1, 2, \{1,2\}\} \in A$
    (h) $\varnothing \subseteq A$

    (i) $\{1\} \subseteq A$
    (j) $\{1,3\} \subseteq A$
    (k) $\{1, 2, \{1,2\}\} \subseteq A$
    (l) $\{\{1,2,3\}\} \in A$

1.5 Determine whether each of the following statements is always true regardless of what sets A and B stand for. Give a short explanation for each answer. If the statement is true, explain why it is true. If not, find a counterexample showing that the statement is not true.

    (a) $A \setminus B = B \setminus A$
    (b) $(A \cup \varnothing) = A$
    (c) $(A \setminus B) \cup B = A$

    (d) $(A \cup B) \cap A = (A \cap B) \cap A$
    (e) $A \setminus (A \cup B) = \varnothing$
    (f) $B \subseteq (A \cup B) \cap A$

1.6 Determine whether each of the following statements is true or false:

    (a) $x \in \{x\}$
    (b) $\{x\} \in \{x\}$
    (c) $\{x\} \subseteq \{x\}$

    (d) $\{x\} \in \{\{x\}\}$
    (e) $\{x\} \subseteq \{\{x\}\}$
    (f) $\varnothing \in \{\varnothing\}$

    (g) $\varnothing \subseteq \{\varnothing\}$
    (h) $\varnothing \in \{\{\varnothing\}\}$
    (i) $\varnothing \subseteq \{\{\varnothing\}\}$

1.7 How many subsets do the sets $\{1,2,3,4\}$ and $\{1,2,3,4,5\}$ have, respectively?

1.8 Find three sets A, B, and C, such that all of the following statements are true:

    – The union of A and B is $\{1,2,3,4\}$.
    – The intersection of A and C is $\{3\}$.
    – The union of B and C is $\{3,4,5,6\}$.
    – The intersection of B and C is $\{4\}$.

1.9 Find two different nonempty subsets A and B of $\{1,2,3\}$ such that $A \cap B = \varnothing$ and $A \cup B = \{1,2,3\}$.

1.10 Show that for all sets X, we have $X \subseteq X$ and $\varnothing \subseteq X$.

1.11 Show that if $A \subseteq B$ and $B \subseteq C$, then $A \subseteq C$.

1.12 Find two sets S and T such that S is both an element of T and a subset of T.

1.13 Show that the following statements are true for all sets A, B, and C:

(a) $(A \setminus B) \setminus C \subseteq (A \setminus C)$

(c) $(A \setminus C) \cap (C \setminus B) = \varnothing$

(b) $(A \cap B) \cap C \subseteq (A \cap B)$

(d) $A \cap (B \cup C) = (A \cap B) \cup (A \cap C)$

1.14 Let $C = A \setminus B$, $D = B \setminus A$, and $E = A \cup B$. Find an expression for $A \cap B$ that does not use any letters other than $C, D,$ and E.

1.15 Let $R = \{ \langle 1, 1 \rangle, \langle 1, 2 \rangle, \langle 1, 3 \rangle, \langle 2, 2 \rangle, \langle 2, 3 \rangle, \langle 3, 3 \rangle \}$ and $S = \{ \langle 3, 1 \rangle, \langle 2, 2 \rangle, \langle 3, 2 \rangle, \langle 3, 3 \rangle \}$.

(a) Calculate $R \cap S$.

(c) Calculate $S \setminus R$.

(b) Calculate $R \setminus S$.

(d) How many subsets does S have?

1.16 How many elements are there in a Cartesian product?

1.17 What is the *largest* set you can imagine? Are there any limits to how large a set can be? And how about the opposite: are there any limits to how *small* a set can be? Can you imagine a set smaller than the empty set?

1.18 Explain why there are two different ways to take the union of multisets. What are the results of the following unions for the multisets $[1, 2]$ and $[1, 2]$: $[1, 2] \cup [1, 2]$ and $[1, 2] + [1, 2]$?

1.19 Find two examples of situations in which it is *better* to use multisets than ordinary sets, as well as two examples in which it is better to use ordinary sets.

1.20 In an attempt to turn *everything* into sets, we define an ordered pair $\langle x, y \rangle$ as the set $\{x, \{x, y\}\}$.

(a) How are $\langle 1, 2 \rangle$ and $\langle 2, 1 \rangle$ to be represented according to this definition?

(b) Will this definition work? In other words, does this definition adequately capture what we mean by an *ordered pair*?

(c) What are the advantages and disadvantages of such a definition?

Chapter 2

Propositional Logic

■░■■■■■■■■■■■■■■■■■■■■

In this chapter you will learn about the concept of a proposition, to represent statements in a proper way using propositional formulas, and to understand the connection between propositions (atomic propositions combined by means of logical words) and formulas (propositional variables that are combined using connectives). You will also learn about necessary and sufficient conditions, and some practical abbreviations.

What Follows from What?

We use logic and logical reasoning all the time. When we talk with each other, when we plan our future, or when we discuss whether something is true, we use logic. The most basic question in logic is what *follows logically* when a set of assumptions is given: If you assume that something is true, what follows logically from that? And what does *not* follow? We will now look at *classical propositional logic*, which is one way to answer these questions.

What Is a Proposition?

Propositional logic is about propositions and which propositions follow from which. In order to speak as precisely as possible, we have to define what we mean by a proposition.

> **Definition 2.1. Proposition**
>
> A **proposition** is something that can be true or false. This something can be a *sentence*, an *utterance*, or the *content of meaning* of these.

Q **Proposition.** Here are some examples of propositions:

- $2 + 2 = 8$.
- *There are infinitely many primes.*
- *Voltage is measured in volts.*
- *This statement is on page 19.*

These fulfill the definition of a proposition because they can be true or false.

The following sentences/expressions are *not* propositions:

© Springer Nature Switzerland AG 2021
R. Antonsen, *Logical Methods*, https://doi.org/10.1007/978-3-030-63777-4_3

– *When is the exam?*	– *Sun and moon.*
– *Hurray!*	– *May curiosity flourish.*

They do not meet the definition of a proposition because they cannot be true or false.   ♦

The definition of a proposition is intentionally a bit vague, and we are not going to delve deeply into the philosophical analysis of what a proposition is. In any case, here are some additional comments for those who are interested:

- A proposition is the underlying content of a statement that can be true or false and that have exactly one of these truth values. Propositions are therefore often called *truth-bearers*.
- A proposition is something about which we can have knowledge and perceptions. We know that "the Earth is orbiting the Sun," and what we then know is an example of a proposition.
- A proposition is in many ways independent of language and form; the sentences "it is raining" and "det regner" (meaning "it is raining" in Norwegian) express exactly the same proposition, although they look different.
- A proposition is the content of a sentence when it is uttered in a given context and when it is unambiguously either true or false.

The important thing when we consider something as a proposition is that we disregard everything but the property that it can be true or false.

Atomic and Composite Propositions

We may analyze and divide propositions into smaller and smaller pieces until we arrive at so-called *atomic* propositions. For example, here are some possible atomic propositions:

– *Oslo is the capital of Sweden.*	– *I'm happy.*
– *All even numbers are divisible by 2.*	– $5 + 6 = 7$.

The internal structure of atomic propositions is not analyzed further, and we can choose our atomic propositions quite freely. However, when we come to first-order logic, we will analyze propositions in greater detail. From atomic propositions we can build *composite* propositions using **logical words**, such as *and*, *or*, *not*, and *if*, *then*. Some composite propositions are, for example, the following:

- *Oslo is the capital of Sweden,* **and** $5 + 6 = 7$.
- **If** $5 + 6 = 7$, **then** *all even numbers are divisible by 2.*
- *I'm happy,* **or** *I'm not happy.*

Among other things, we will look at the following questions:

- How does the truth value of a composite proposition depend on the truth values of the atomic propositions it is made up of?

– What propositions are true *independent* of the truth values of the atomic propositions? Such propositions are called *tautologies*.

We shall now *abstract* over propositions and *represent* them symbolically. We will use symbols and signs to represent and calculate with propositions just as we do for other mathematical objects.

Atomic and Composite Formulas

In the same way as composite propositions are made up of atomic statements and logical words, we will create a symbolic language in which smaller expressions are combined into larger expressions representing composite propositions. To represent atomic propositions we use so-called propositional variables.

Definition 2.2. Propositional variable

A **propositional variable** is a variable, $P, Q, R, \ldots$ or something similar. A propositional variable is an **atomic formula**.

Exactly what symbols we use as propositional variables is not so important. What matters is that we have *enough* propositional variables to express and represent everything we want. Here, we use capital letters. From now on we assume that an adequate set of propositional variables is given.

Just as variables in mathematics can be used represent numbers, like x and y in the expression $x + 2 \cdot y = 2 \cdot y + x$, and letters in set theory represent sets, like A and B in the expression $A \cup B = B \cup A$, propositional variables represent propositions. The propositions are *concrete*, and the variables are *abstract*. The propositional variables are the symbols we use to represent the propositions.

To capture and represent composite propositions, for example the proposition "**if** you lift weights, **then** you become strong," we need more symbols in the language. In this case we have two natural *atomic* propositions, namely "you lift weights" and "you become strong." These propositions may be represented by the propositional variables T and S. The composite proposition may now be represented by means of a propositional formula $(T \rightarrow S)$. The symbols we are going to use for constructing composite expressions are called *connectives*.

Definition 2.3. Connective

The **logical connectives** are $\neg, \wedge, \vee$, and $\rightarrow$.

The intuition behind the connectives is that they represent the logical words: $\neg$ means *not*, $\wedge$ means *and*, $\vee$ means *or*, and $\rightarrow$ means *implies* or *if, then*.

Digression

There are numerous other connectives, but these four are the most common. We could also have defined ← (*if* / *converse implication*), ≡ or ↔ (*if and only if*), ↑ or | (*NAND* / *Sheffer stroke* / *not and*), ↓ (*NOR* / *neither nor*), ↛ or ⊕ (*XOR* / *exclusive or*), ↛ (*does not imply*), and many more. One reason why these are not defined is that they can be defined *in terms of* ¬, ∧, ∨, and → in a natural way. In fact, one could get by with only ¬ and ∨, or even with only |.

The connectives are part of the symbolic language; they are the *symbols* we are going to use to combine smaller formulas into larger ones. The following definition precisely defines what a propositional formula is.

Definition 2.4. Propositional formulas

Every atomic formula is a propositional formula. If F and G are propositional formulas, then the following are also propositional formulas:

- ¬F is a propositional formula; we call it the **negation** of F. This represents the proposition "not F."
- (F ∧ G) is a propositional formula; we call it the **conjunction** of F and G. This represents the proposition "F and G." The formulas F and G are called the **conjuncts**.
- (F ∨ G) is a propositional formula; we call it the **disjunction** of F and G. This represents the proposition "F or G." The formulas F and G are called the **disjuncts**.
- (F → G) is a propositional formula; we call it the **implication** from F to G. This represents the proposition "if F, then G."

Only expressions that can be constructed in this way are propositional formulas.

Notice that F and G work like *placeholders* in the definition; both F and G may stand for composite formulas.

Q   Propositional formulas. If P, Q, and R are three propositional variables, we have:

- P, Q, and R are propositional formulas, because they are propositional variables.
- ¬P is a propositional formula, because P is.
- (Q ∧ R) is a propositional formula, because Q and R are.
- (¬P ∨ (Q ∧ R)) is a propositional formula, because ¬P and (Q ∧ R) are.

We can see that the formula (¬P ∨ (Q ∧ R)) is constructed from within, beginning with the propositional variables and working our way outward. We can visualize the construction as a tree in the following way:

It is useful to categorize formulas according to what the *outermost* connective, or the **principal connective**, of the formula is. For example, if F and G are formulas, then (F ∧ G) is a formula, and ∧ is the principal connective of the formula. In this case, we say that the formula is an ∧-formula, and correspondingly for ¬-, ∨- and →-formulas. The formula (¬P ∨ (Q ∧ R)) from the example above is an ∨-formula. We do exactly the same with arithmetic expressions like $-1 + (2 \cdot 3)$, which is a sum and not a product.

Hence, propositional formulas are built up using only propositional variables, connectives, and parentheses. With this starting point, we can represent propositions in a precise way.

Q   Propositional formulas. If S and T are propositional formulas, then (¬S → ¬T) is also a propositional formula. If S represents the proposition *you become strong*, and T represents the proposition *you exercise*, then (¬S → ¬T) represents the proposition *if you are not becoming strong, then you do not exercise.* ◆

Q   Representation. Here is a way to represent propositions using propositional formulas. We assume that P stands for the proposition *pigs like the polka* and that Q stands for the proposition *pigs like disco.*

Pigs like the polka.	P
Pigs like disco.	Q
*Pigs **do not** like disco.*	¬Q
*Pigs like **both** the polka **and** disco.*	(P ∧ Q)
*Pigs **do not** like **both** the polka **and** disco.*	¬(P ∧ Q)
*Pigs like the polka **or** disco.*	(P ∨ Q)
*Pigs like the polka **or do not** like the polka.*	(P ∨ ¬P)
*Pigs like **neither** the polka **nor** disco.*	(¬P ∧ ¬Q) or ¬(P ∨ Q)
*If pigs like the polka, **then** they like disco.*	(P → Q)
*If pigs **do not** like the polka, **then** they like disco.*	(¬P → Q)
*Pigs like the polka **if** they like disco.*	(Q → P)
*Pigs like the polka **only if** they like disco.*	(P → Q)
*It is **not** the case that pigs **do not** like disco.*	¬¬Q.

There are often many different ways to represent a proposition, and it may be useful to use the structure of the proposition as a guideline. ◆

❷   Find a propositional formula that represents the proposition "if I win the lottery or pass the exam, then I am happy." ◆

❶   Let L stand for "I win the lottery," let E stand for "I pass the exam," let H stand for "I am happy." We may then represent the proposition as ((L ∨ E) → H). ◆

Necessary and Sufficient Conditions

The words *necessary* and *sufficient* are often used in logical reasoning, and these words are closely related. The formula $(A \to B)$ expresses "if A, then B." It is worth thinking about what that means. The following means exactly the same as "if A, then B":

  – *The truth of A is a **sufficient** condition for the truth of* B.

This means that it is *enough* for A to be true in order for B to be true as well; hence the word "sufficient." The formula B can be true without A being true, but *if* A is true, *then* B must also be true. We can also formulate exactly the same thing in the following way using the word "necessary":

  – *The truth of B is a **necessary** condition for the truth of* A.

This means that A cannot be true without B also being true; hence the word "necessary." The formula A is true *only if* B is true. All these express the same thing, and we can represent it with the formula $(A \to B)$.

Let us look at the difference between the propositions "I eat something **if** it is tasty" and "I eat something **only if** it is tasty." *Do you see that they each mean something different?*

The first proposition may be analyzed as follows:

  – *I eat something **if** it is tasty.*
  – ***If** something is tasty, **then** I eat it.*
  – *The fact that something is tasty is a **sufficient** condition for me to eat it.*
  – In other words: *I am a glutton who eats everything tasty.*
    (Note that in this case, I may, if I so choose, eat something even if it is not tasty.)

The second proposition may be analyzed as follows:

  – *I eat something **only if** it is tasty.*
  – ***If** I eat something, **then** it is tasty.*
  – *The fact that something is tasty is a **necessary** condition for me to eat it.*
  – In other words: *I am picky, eating only tasty things.*
    (In this case, something may be tasty without my eating it.)

We can also look at the combination of propositions using "if and only if." This may be analyzed as follows:

  – *I eat something **if and only if** it is tasty.*
  – *I eat something **if** it is tasty, and I eat something **only if** it is tasty.*
  – In other words: *I am a glutton, but I am also picky.*

It is quite common to use the word "iff," with two f's, as an abbreviation for "if and only if": "I eat something **iff** it is tasty."

We will also use the following practical abbreviations for formulas representing "if and only if" propositions.

Notation. If F and G are propositional formulas, then $(F \leftrightarrow G)$ stands for the formula $(F \rightarrow G) \land (G \rightarrow F)$.

❷ Find a propositional formula that represents the proposition "I am happy **if and only if** I win the lottery." ◆

❶ Let H stand for "I am happy," and let L stand for "I win the lottery." Now we can split the proposition into two parts:

*I am happy **if** I win the lottery*:	$(L \rightarrow H)$
*I am happy **only if** I win the lottery*:	$(H \rightarrow L)$

We can represent the entire proposition with $(H \rightarrow L) \land (L \rightarrow H)$ or $(H \leftrightarrow L)$. ◆

Note. The word "if" is used in a very particular way in *definitions*. It is not used as part of a *claim* that something is true, but rather to *define* the meaning of a word. For example, look at the definition "a word is a **palindrome** if it reads the same backward as forward." Here the meaning of the word "palindrome" is given, and it is not a *claim* that something is a palindrome given that a condition is satisfied. If it had been such a claim, then something could have been a palindrome even though the condition was *not* satisfied, because the claim specifies only a *sufficient* and not a *necessary* condition for something to be a palindrome. Occasionally, "if and only if" is used for definitions, but it is unnecessary, provided that we understand that in definitions, the "only if" part is always understood.

Parentheses, Precedence Rules, and Practical Abbreviations

The set of propositional formulas is precisely defined, and expressions like $P \lor Q$ and $P \lor Q \rightarrow R$ are not, strictly speaking, elements of this set, since the parentheses are missing. We shall nevertheless accept such expressions as propositional formulas. But then we also need some conventions for removing ambiguities.

Definition 2.5. Precedence rules for the connectives

We give the connectives different **precedences** in relationship to each other:

– $\neg$ binds most strongly.   – $\lor$ binds more weakly than both $\neg$ and $\land$.

– $\land$ binds more weakly than $\neg$.   – $\rightarrow$ binds the most weakly.

That $\bullet$ binds more strongly than $\circ$ means that $P \bullet Q \circ R$ stands for $((P \bullet Q) \circ R)$ and not $(P \bullet (Q \circ R))$. Additionally, $\land$ and $\lor$ are **left-associative**, and $\rightarrow$ is **right-associative**. That $\bullet$ is *left*-associative means that $P \bullet Q \bullet R$ stands for $((P \bullet Q) \bullet R)$ and not $(P \bullet (Q \bullet R))$, and correspondingly for right-associative.

This is completely analogous to what we do with numbers, where it is common practice that multiplication binds more strongly than addition: The expression $2 \cdot 3 + 4$ means $(2 \cdot 3) + 4$ and not $2 \cdot (3 + 4)$.

Q **Precedence rules.** The precedence rules give us the following:

- The expression $\neg P \rightarrow Q$ stands for $(\neg P \rightarrow Q)$ and *not* $\neg(P \rightarrow Q)$.
  That is because $\neg$ binds more strongly than $\rightarrow$.
- The expression $P \rightarrow Q \wedge R$ stands for $(P \rightarrow (Q \wedge R))$ and *not* $((P \rightarrow Q) \wedge R)$.
  That is because $\wedge$ binds more strongly than $\rightarrow$.
- The expression $P \wedge Q \wedge R \rightarrow S \vee T \vee U$ stands for $((P \wedge Q) \wedge R) \rightarrow ((S \vee T) \vee U)$.
  That is because $\wedge$ and $\vee$ are left-associative and bind more strongly than $\rightarrow$.
- The expression $P \rightarrow Q \rightarrow R$ stands for $(P \rightarrow (Q \rightarrow R))$ and *not* $((P \rightarrow Q) \rightarrow R)$.
  That is because $\rightarrow$ is right-associative. ◆

Defining precedence rules is important for several reasons. First and foremost, it makes complex expressions easier to read and write, but it is also an important clarification of the *meaning* of the connectives. To create a computer program that calculates the value of the expression $1 + 2 + 3 + 4 + 5$, you must specify the *order* in which the numbers are to be added. This is also the case for propositional logic.

Let us look at the propositional formula $P \rightarrow Q \rightarrow R$. The rules state that $\rightarrow$ is *right-associative*, which means that the formula actually stands for $(P \rightarrow (Q \rightarrow R))$. But why is $\rightarrow$ right-associative by convention and not left-associative? That is because it is natural to read the formula from left to right as "if P, then ..." and then "if Q, then ...." If $\rightarrow$ had been left-associative, the formula would have said $(P \rightarrow Q) \rightarrow R$, but we rarely need to say this.

> **Digression**
>
> A widely used technique in mathematics and computer science is called **currying**, named after the American mathematician and logician *Haskell B. Curry* (1900–1982), after whom the functional programming language *Haskell* is also named. In the context of propositional formulas, currying means to change expressions of the form $(F \wedge G) \rightarrow H$ into expressions of the form $F \rightarrow (G \rightarrow H)$. In the first formula there is a *composite* formula on the left side of $\rightarrow$, while in the second formula there is a simpler formula there. One can look at this a bit like "unpacking" what is in $F \wedge G$. This technique is used, among other things, for changing functions with multiple arguments into sequences of functions such that each function takes only a single argument. We will talk about functions in Chapter 7, when this will become more meaningful.

Exercises

2.1 Let H stand for "I am happy," let K stand for "you kiss me," and let E stand for "I pass the exam." Write down the propositions represented by the following propositional formulas:

(a) $(K \to H)$

(b) $(K \vee \neg K)$

(c) $\neg(K \wedge \neg H)$

(d) $\neg(K \vee E)$

(e) $((\neg K \wedge \neg E) \to \neg H)$

(f) $((E \to K) \wedge (\neg E \to K))$

2.2 Find propositional formulas that represent the following propositions:

(a) *I will be happy if you come to the party.*

(b) *I will be happy only if you come to the party.*

(c) *I will be happy if and only if you come to the party.*

(d) P *is necessary for* Q.

(e) P *is sufficient for* Q.

(f) P *is necessary and sufficient for* Q.

2.3 Let D stand for the proposition "the penguins are dancing," let S stand for the proposition "the penguins are singing," and let G stand for the proposition "the penguins are happy." Represent the following propositions using the propositional variables D, S, and G and the connectives $\wedge$, $\vee$, $\to$, and $\neg$:

(a) *The penguins are singing and dancing.*

(b) *The penguins are not happy.*

(c) *If the penguins are happy, they are dancing or singing.*

(d) *If the penguins are not happy, they are neither dancing nor singing.*

(e) *The penguins are singing and happy, but not dancing.*

(f) *The penguins are singing and dancing, or they are not happy.*

(g) *If the penguins are dancing, they are also happy and singing.*

(h) *If the penguins are not dancing, they are not happy.*

2.4 Let the propositional variable P stand for "I like Piglet," let T stand for "I am tough," and let R stand for "I enjoy reading." Find the propositions represented by the following propositional formulas:

(a) $(P \wedge T)$

(b) $((R \wedge T) \to P)$

(c) $\neg(P \to \neg T)$

Find propositional formulas that represent the following propositions:

(d) *I do not like Piglet, nor am I tough, but I enjoy reading.*

(e) *If I enjoy reading, I am both tough and I like Piglet.*

2.5   Explain the difference between the following propositions:

(a) "I promise to tell the truth."

(b) "I promise to tell the truth and only the truth."

2.6   What is the difference between the following pairs of propositions? Do they mean the same, do they mean the opposite of each other, or neither?

(a) *If I solve the problem, then I am smart.*
If I am smart, then I solve the problem.

(b) *If x is an element of A ∪ B, then x is an element of A or an element of B.*
If x is an element of neither A nor B, then x is not an element of A ∪ B.

(c) *The book is not both entertaining and educational.*
The book is not entertaining, or the book is not educational.

(d) *If I am happy, then I am satisfied, and if I am satisfied, then I am happy.*
I am both satisfied and happy.

2.7   Use the precedence rules to put all the parentheses in the right places:

(a) $P \land Q \rightarrow R \lor S$

(b) $P \lor Q \land R \lor S$

(c) $P \land Q \land R \rightarrow S$

(d) $P \rightarrow Q \land R \rightarrow S \rightarrow T$

2.8   "This sentence has five words."

(a) What is the truth value of this statement?

(b) Write down the negation of the statement. What is the truth value of the negation?

2.9   Look at the statement $\{1, 2\} \subseteq A$.

(a) Find a condition for A that is *sufficient*, but *not necessary*, for the statement to be true.

(b) Find a condition for A that is *necessary*, but *not sufficient*, for the statement to be true.

(c) Find a condition that is both sufficient and necessary.

(d) Find a condition that is neither sufficient nor necessary.

Chapter 3

Semantics for Propositional Logic

In this chapter you will learn semantics for propositional logic. That is, you will learn how to systematically give interpretations to propositional formulas. An assignment of truth values to propositional variables gives rise to a valuation, and this provides truth values for all formulas. You will learn about truth tables, a tool for exploring valuations, about the concept of logical equivalence, and some known logical equivalences.

Interpretation of Formulas

We have so far learned a formal language for representing propositions; we have defined the *syntax* of propositional logic. Now we will look at how we can give the formal language an *interpretation*; we are going to define the *semantics* of propositional logic. We are going to interpret propositional formulas as *true* or *false*. The following definition gives us a practical abbreviation for the truth values *true* and *false*.

Definition 3.1. Truth values

We let **1** and **0** stand for the **truth values true** and **false**.

It may be useful to compare the situation here with the corresponding situation for numbers. If someone asks us what the value of $(x + y)$ is, it is impossible to answer without knowing the values of both x and y. The situation is just like this for propositional logic: In order to know the value a propositional formula, say $(P \vee Q)$, we have to know the values of the propositional variables P and Q. We shall interpret each propositional variable as either true or false, and from this, the truth values of all other propositional formulas will follow. The next definition makes this precise.

© Springer Nature Switzerland AG 2021
R. Antonsen, *Logical Methods*, https://doi.org/10.1007/978-3-030-63777-4_4

Definition 3.2. Interpretation of propositional formulas

If the truth values of the propositional formulas F and G are given, then the truth values of ¬F, (F ∧ G), (F ∨ G), and (F → G) are given by the following tables, which show how the connectives are to be interpreted:

F	¬F
1	0
0	1

F	G	(F ∧ G)
1	1	1
1	0	0
0	1	0
0	0	0

F	G	(F ∨ G)
1	1	1
1	0	1
0	1	1
0	0	0

F	G	(F → G)
1	1	1
1	0	0
0	1	1
0	0	1

Notice that F and G are *placeholders* in the definitions; both F and G may stand for composite formulas.

Q **Negation.** If F stands for the proposition "it is raining outside," then ¬F stands for the proposition "it is **not** raining outside." The truth value of ¬F depends directly on the truth value of F. If F is true, then ¬F is false, and vice versa.   ♦

Q **Conjunction.** If F stands for the proposition "the ball is small," and G stands for the proposition "the ball is red," then (F ∧ G) stands for the proposition "the ball is small **and** the ball is red." If this proposition is to be true, then the ball in question must be *both* small and red. This is captured in the top row of the table. If the ball is not small or not red, the conjunction is false.   ♦

Q **Disjunction.** If we continue with the example of the ball, (F ∨ G) stands for the proposition "the ball is small **or** the ball is red." If this proposition is to be true, then the ball in question must be small *or* red. The only way for the disjunction to be false is by having a ball that is neither small nor red. This is captured in the last row of the table. Notice that (F ∨ G) is also true when both F and G are true; for this reason we call ∨ the *inclusive or*.   ♦

Q **Implication.** If F stands for the proposition "the woman speaks French," and G stands for "the woman speaks Greek," then (F → G) stands for the implication "**if** the woman speaks French, **then** the woman speaks Greek." If we are dealing with a woman who speaks French but does not speak Greek, then the implication is *false*; this is captured in the second row of the table. This is also the only way to make an implication false. If the woman speaks both languages, which is captured in the top row of the table, the implication is true, and if the woman does not speak French, which is captured in the two last rows of the table, it is irrelevant whether she speaks Greek for the implication to be true.   ♦

Valuations and Truth Tables

We often talk about an *assignment of truth values* to propositional variables. That means that for each propositional variable, we have chosen a specific value, either **1** or **0**. When such an assignment is chosen, there is exactly one way to assign a value to *every* propositional formula, given that we are interpreting the connectives as in Definition 3.2 (*page 30*).

Definition 3.3. Valuation

An assignment of truth values to all propositional formulas that is in agreement with the tables in Definition 3.2 is called a **valuation**.

When we provide a valuation, we only need to provide values for the propositional variables. Then the values for all other propositional formulas are determined as well.

Q **Valuation.** Look at the formula $(P \wedge \neg Q)$, and suppose that we have a valuation that makes P true and Q false. This valuation must make $\neg Q$ true, because of how $\neg$-formulas are interpreted. We have then shown that both P and $\neg Q$ are true, and we can conclude that $(P \wedge \neg Q)$ is true, because of how $\wedge$-formulas are interpreted. ♦

A practical way to compute valuations is using truth tables.

Definition 3.4. Truth table

A **truth table** is a table that tells us the truth value of a composite propositional formula, based on which truth values are assigned to the propositional variables.

A truth table consists of vertical columns and horizontal rows. It is common to have a *column* for each propositional variable to the left, while each *row* in the table represents a specific assignment of truth values to the propositional variables. We can construct a truth table for the formula $(P \wedge \neg Q)$ as follows:

P	Q	(	P	$\wedge$	$\neg$	Q	)
1	1		1	0	0	1	
1	0		1	1	1	0	
0	1		0	0	0	1	
0	0		0	0	1	0	

The truth values for the formula are indicated below the $\wedge$ symbol in boldface. We see, in the second row, that there is only one valuation that makes the formula true, and that is the one that makes P true and Q false. We see that it corresponds to what the formula expresses; namely "P and not Q."

❷ Suppose we have a valuation that makes both P and Q false. Will this valuation make the formula $(\neg P \rightarrow Q)$ true or false? ◆

❗ We have assumed that P is false, and therefore $\neg P$ must be true, because of how $\neg$-formulas are interpreted. We have also assumed that Q is false, and therefore $(\neg P \rightarrow Q)$ must be false, because of how $\rightarrow$-formulas are interpreted. We can also construct a truth table and find the answer as follows:

P	Q	(	¬	P	→	Q	)
1	1		0	1	1	1	
1	0		0	1	1	0	
0	1		1	0	1	1	
0	0		1	0	0	0	

We find the valuation that makes P and Q false in the last row. Here we see that the whole formula, $(\neg P \rightarrow Q)$, is false. ◆

Properties of Implication

A formula $F \rightarrow G$ is true in all cases except when F is true and G is false, and there are several reasons that it is beneficial to define the interpretation of $\rightarrow$-formulas in this way.

The proposition "If you have a ticket, then you may enter" can be represented by the propositional formula $T \rightarrow E$, where T represents "you have a ticket" and E represents "you may enter." Think about what it means to know that this proposition is (1) *true* and (2) *false*, respectively.

(1) *What do you know if you know that the proposition is true?* Then, E must be true – that you may enter – *given* that T is true – that you have a ticket. It is a kind of guarantee that says what must be the case if T is true. What happens when T is false – that you do not have a ticket – does not matter for the truth value of the entire proposition; then E – that you may enter – can be true or false.

(2) *What do you know if you know that the proposition is false?* Then T must be true – that you have a ticket – and E must be false – that you may not enter. This is the *only* way to make the whole proposition false.

In order to motivate the truth table for an implication, ask yourself the following questions: *What is needed to confirm that the proposition is true?* Imagine that "if you have a ticket, you may enter" is a theory whose truth value you want to determine. What will confirm or refute this theory? What certainly confirms the theory is that you may actually enter if you have a ticket. If you do not have a ticket, this is also a type of confirmation, because it does not provide a *counterexample*. The only thing that can refute the theory is that you have a ticket, but may not enter. For example, it is not natural to say that it is a refutation if you do not have a ticket, but may enter anyway. *If you know that the proposition is true, what do you know then? What information is inherent in the fact that the proposition is true?* If you get new information, and this is in the form of an "if, then" statement, what information have you then received?

Suppose that I say "if you come to the party, I will be happy." *What is the case if this sentence is not true?* It must be that you came to the party, but I didn't become happy; this is the only scenario that would show that the proposition was false.

When we talk about propositional logic and valuations, it is only *truth*, and how the truth values of composite propositions depend on the truth values of propositional variables, that is our focus. Many other aspects, for example time, "he brushed his teeth and went to bed," and physical causality, "if you blow the trumpet, there will be sound," get lost in the representation.

Logical Equivalence

The notion of *logical equivalence* is one of the most important concepts in logic, because it says in a precise way what it means for two formulas to have the same meaning.

Definition 3.5. Logical equivalence

Two formulas F and G are **logically equivalent**, or just **equivalent**, if they have the same truth value for each assignment of truth values to the propositional variables. Put another way, all valuations that make F true must make G true, and vice versa. We write F $\Leftrightarrow$ G when F and G are logically equivalent.

Q **Equivalent formulas.** The formulas P and $\neg\neg$P are equivalent. If a valuation makes P true, it must also make $\neg\neg$P true; if a valuation makes $\neg\neg$P true, it must also make P true. They always have the same truth value.   ♦

If two formulas are equivalent, they have the same meaning content. If two formulas are *not* equivalent, we can find a valuation that makes one formula true and the other formula false.

Q **Inequivalent formulas.** The formulas P $\vee$ Q and P $\wedge$ Q are not equivalent. A valuation that makes P true and Q false will make P $\vee$ Q true, but P $\wedge$ Q false. They do not always have the same truth value and are therefore not equivalent.   ♦

We use the following expressions interchangeably:

- F and G are equivalent.
- F $\Leftrightarrow$ G
- F is true if and only if G is true.
- F is true iff G is true.

   – F is a necessary and sufficient condition for G.

Note. It is useful to distinguish between the *metalanguage*, which is what we use when we are reading, writing, and speaking, and the *object language*, which in this case is the formal propositional language. Instead of "if and only if" we often write ⇔, but this symbol is not part of the *formal* language. It is used by us – in the *metalanguage* – as an abbreviation for "if and only if."

Q **Equivalent formulas.** The propositional formulas $(P \rightarrow Q)$ and $(\neg P \vee Q)$ are equivalent. This is easiest to see using a truth table:

P	Q	(P	→	Q)	(¬P	∨	Q)
1	1	1	1	1	0	1	1
1	0	1	0	0	0	0	0
0	1	0	1	1	1	1	1
0	0	0	1	0	1	1	0

We see that the formulas have the same truth value regardless of the assignment of truth values to the propositional variables.   ♦

It is not necessary to use truth tables in order to argue that two formulas are equivalent. It is also possible to use a completely ordinary argument as well, and this is often much better.

? Show that the formulas $\neg(P \rightarrow Q)$ and $(P \wedge \neg Q)$ are equivalent.   ♦

! To prove this, it is enough to show that $\neg(P \rightarrow Q)$ is true if and only if $(P \wedge \neg Q)$ is true. First suppose that $\neg(P \rightarrow Q)$ is true.

   – Then $(P \rightarrow Q)$ must be false, because of how $\neg$-formulas are interpreted.
   – Then P must be true and Q false, because of how $\rightarrow$-formulas are interpreted.
   – Then $\neg Q$ must be true, because of how $\neg$-formulas are interpreted.
   – Then $(P \wedge \neg Q)$ must be true, because of how $\wedge$-formulas are interpreted.

Furthermore, suppose that $(P \wedge \neg Q)$ is true.

   – Then P and $\neg Q$ must be true, because of how $\wedge$-formulas are interpreted.
   – Then Q must be false, because of how $\neg$-formulas are interpreted.
   – Then $(P \rightarrow Q)$ must be false, because of how $\rightarrow$-formulas are interpreted.
   – Then $\neg(P \rightarrow Q)$ must be true, because of how $\neg$-formulas are interpreted.   ♦

? We define a new connective, ↓, which is often called *NOR*, and interpret $(P \downarrow Q)$ as follows:

P	Q	$(P \downarrow Q)$
1	1	0
1	0	0
0	1	0
0	0	1

(a) Find a formula that is logically equivalent to $(P \downarrow Q)$ in which the only connectives are $\neg$ and $\wedge$. Explain your answer by creating a truth table.

(b) Find a formula that is logically equivalent to $\neg P$ in which the only connective is $\downarrow$. Explain your answer by creating a truth table. ◆

♨ (a) The formula $(\neg P \wedge \neg Q)$ is equivalent to $(P \downarrow Q)$ and uses only the connectives $\neg$ and $\wedge$. We can see this in the following truth table, since the two columns under the principal connectives of the formulas are identical:

P	Q	(P	↓	Q)	(¬P	∧	¬Q)
1	1	1	0	1	0	0	0
1	0	1	0	0	0	0	1
0	1	0	0	1	1	0	0
0	0	0	1	0	1	1	1

(b) The formula $(P \downarrow P)$ is logically equivalent to $\neg P$ and uses only the connective $\downarrow$. We can see this in the following truth table:

P	¬P	(P ↓ P)
1	0	0
0	1	1

◆

Digression

What we are studying here is called *classical* propositional logic. What characterizes classical propositional logic is that $P \vee \neg P$ is always true, and that $\neg \neg P$ is equivalent to P. However, it is possible to define logics that behave differently. In **intuitionistic logic** and **constructive logic** it is *not* the case that $P \vee \neg P$ is always true or that $\neg \neg P$ is equivalent to P. In these logics, the focus is on *provability*. The interpretations of the connectives are formulated as conditions of provability: We say that the formula $F \vee G$ is *provable* in intuitionistic logic if F is provable or G is provable. And this is why $P \vee \neg P$ is not valid in intuitionistic logic, because by definition neither P nor $\neg P$ is provable. In intuitionistic logic we must therefore define *another* semantics; ordinary truth tables and valuations are no longer sufficient. We will meet intuitionistic logic again in Chapter 24.

A Study in What Is Equivalent

Finding exactly which formulas are logically equivalent to each other is one of the most essential things we do in logic. And knowing these equivalences may also be useful in programming, for example if we want to understand and simplify complex conditions. Now we will look at some of the most famous and useful equivalences. Many of these appear so often that they are called *laws*. There are more than these, and this is just a taste:

Distributive laws describe how conjunction and disjunction *distribute* over each other in the following way:

$A \wedge (B \vee C) \Leftrightarrow (A \wedge B) \vee (A \wedge C)$,
$A \vee (B \wedge C) \Leftrightarrow (A \vee B) \wedge (A \vee C)$.

De Morgan's laws, named after the British mathematician *Augustus De Morgan* (1806–1871), describe the interaction between negation on the one hand, and conjunction and disjunction on the other. Intuitively, these say that the negation sign can be "pushed inward" in a formula, but that we then have to change the connectives:

$\neg(A \wedge B) \Leftrightarrow (\neg A \vee \neg B)$;
if not both A *and* B *are true, then* A *is false or* B *is false.*
$\neg(A \vee B) \Leftrightarrow (\neg A \wedge \neg B)$;
if neither A *nor* B *is true, then* A *is false and* B *is false.*

Associative laws indicate that parentheses may be freely placed when we have only conjunctions or only disjunctions in a formula. That is why it is OK to drop parentheses around them:

$A \wedge (B \wedge C) \Leftrightarrow (A \wedge B) \wedge C$,
$A \vee (B \vee C) \Leftrightarrow (A \vee B) \vee C$.

Commutative laws say that the order of the formulas does not affect the truth value of conjunctions and disjunctions:

$A \wedge B \Leftrightarrow B \wedge A$,
$A \vee B \Leftrightarrow B \vee A$.

We also have the law of **double negation**, which says that two negation signs next to each other may be removed:

$\neg\neg A \Leftrightarrow A$.

We also have the following equivalences for →-formulas:

$A \rightarrow B \Leftrightarrow \neg A \vee B$,
$\neg(A \rightarrow B) \Leftrightarrow A \wedge \neg B$.

Exercises

3.1 (a) Construct the complete truth table for $(P \to Q) \to P$.

(b) Construct the complete truth table for $(P \to Q) \vee (Q \to P)$.

(c) What is special about the truth values of these formulas?

3.2 Construct the truth tables for the following formulas:

(a) $\neg(P \vee \neg Q) \vee P$ (b) $((P \wedge R) \vee (Q \wedge R)) \to (P \to \neg Q)$

3.3 Suppose that A and B are true, and that C and D are false. Find the truth values of the following formulas:

(a) $\neg\neg A$ (f) $(A \vee B)$ (k) $(C \to D)$

(b) $(A \wedge B)$ (g) $(C \vee \neg D)$ (l) $((B \vee C) \to A)$

(c) $(A \wedge C)$ (h) $(A \to B)$ (m) $((B \to C) \to A)$

(d) $(A \wedge \neg C)$ (i) $(A \to C)$ (n) $((C \to B) \to D)$

(e) $(\neg A \wedge B)$ (j) $(C \to A)$ (o) $((A \vee \neg B) \wedge D)$

3.4 For each of the following propositional formulas, find truth values of P, Q, and R such that the formula becomes true:

(a) P (g) $(P \wedge (\neg Q \wedge \neg R))$ (m) $(P \to \neg Q)$

(b) $\neg P$ (h) $(P \vee Q)$ (n) $(\neg P \to \neg Q)$

(c) $\neg\neg P$ (i) $(P \vee \neg Q)$ (o) $(P \to (\neg Q \to \neg R))$

(d) $(P \wedge Q)$ (j) $(\neg P \vee \neg Q)$ (p) $(P \wedge (Q \to R))$

(e) $(P \wedge \neg Q)$ (k) $(P \vee (\neg Q \vee \neg R))$ (q) $(P \vee (P \wedge \neg P))$

(f) $(\neg P \wedge \neg Q)$ (l) $(P \to Q)$ (r) $((P \wedge Q) \vee \neg R)$

3.5 For each of the formulas in the previous exercise, find truth values of P, Q, and R such that the formula becomes false.

3.6 For each of the sets below, determine whether it is possible to make the formulas in the set true simultaneously. Justify your answer. For example, if the set is $\{P, \neg Q\}$, the answer is yes, because if P is true and Q is false, then both P and $\neg Q$ are true formulas.

(a) $\{P, Q, \neg R\}$ (c) $\{P \vee Q, \neg P \vee \neg Q\}$

(b) $\{P, \neg P\}$ (d) $\{P \wedge Q, \neg P \vee \neg Q\}$

3.7 For each of the statements below, determine whether it is true or false. Justify your answer and explain why it is so.

(a) *The formula $(P \wedge \neg P)$ is false for all valuations.*

(b) *It is possible to make the formula $(P \vee \neg P)$ false.*

(c) *If the formula $(P \vee (P \wedge Q))$ is true, then P is true.*

(d) *If the formula P is false, then* $(P \lor (P \land Q))$ *is false.*

(e) *If the formula ¬P is false, then ¬¬P is true.*

(f) *Every propositional formula can be made true and can also be made false.*

3.8   The following truth table shows how the truth values of two formulas F and G depend on the truth values of the propositional variables P, Q, and R:

P	Q	R	F	G
1	1	1	1	0
1	1	0	0	0
1	0	1	0	1
1	0	0	0	0
0	1	1	1	0
0	1	0	1	1
0	0	1	1	0
0	0	0	1	1

(a) Find a formula F with the propositional variables P, Q, and R such that F receives the truth values given in the table, depending on the truth values of P, Q, and R.

(b) Do the same for G.

(c) Find a method for solving *all* exercises of this form.

3.9   Construct the truth table for the formula $(P \leftrightarrow Q)$.

3.10   What are the advantages and disadvantages of truth tables? What are the pros and cons of reasoning without them?

3.11   For each of the following formulas, find an equivalent formula with as few connectives as possible:

(a) $\neg(\neg\neg P \land \neg Q)$

(b) $\neg(\neg P \lor (\neg Q \to \neg R))$

(c) $(\neg P \land (Q \to P))$

(d) $(Q \land (\neg\neg Q \to \neg Q))$

3.12   Show that the following pairs of formulas are *not* equivalent.

(a) $(P \lor Q)$ and $(P \land Q)$

(b) $(P \lor Q)$ and $(P \to Q)$

(c) $(P \to Q)$ and $(Q \to P)$

(d) $(P \to Q)$ and $(\neg P \to \neg Q)$

(e) $(P \to Q)$ and $((P \land R) \to Q)$

(f) $(P \to (Q \lor R))$ and $(P \to (Q \land R))$

3.13   Group the formulas such that all the equivalent formulas end up in the same group:

$(A \land B)$	$\neg(A \lor B)$	$\neg(A \to B)$
$(A \land \neg B)$	$\neg(A \lor \neg B)$	$\neg(A \to \neg B)$
$(\neg A \land B)$	$\neg(\neg A \lor B)$	$\neg(\neg A \to B)$
$(\neg A \land \neg B)$	$\neg(\neg A \lor \neg B)$	$\neg(\neg A \to \neg B)$

3.14   Let A, B, and C be propositional formulas. Show with truth tables that the following equivalences hold:

(a) $A \to B \Leftrightarrow \neg A \lor B$

(b) $\neg(A \to B) \Leftrightarrow A \land \neg B$

(c) $A \lor (B \land C) \Leftrightarrow (A \lor B) \land (A \lor C)$

3.15 Let A, B, and C be propositional formulas.

  (a) Show that $(A \leftrightarrow B)$ is equivalent to $(B \leftrightarrow A)$.
  (b) Show that $(A \leftrightarrow B) \leftrightarrow C$ is equivalent to $A \leftrightarrow (B \leftrightarrow C)$.
  (c) Explain why this means that the placement of parentheses does not matter in expressions that use only the connective $\leftrightarrow$.

3.16 Suppose that you are visited by an alien from a strange planet that says, "I understand well what $\neg$ and $\wedge$ mean, but I do not understand what $\vee$ means. Can you explain it by referring only to $\neg$ and $\wedge$, which I already understand?" What our guest wants is a way to define $\vee$ in terms of $\neg$ and $\wedge$. Find such a formula, that is, a formula with the propositional variables P and Q and the connectives $\neg$ and $\wedge$ such that this formula is equivalent to $(P \vee Q)$. Can you do the opposite as well, that is, find a formula that is equivalent to $(P \wedge Q)$, but that contains only $\neg$ and $\vee$?

3.17 Can $\wedge$ and $\vee$ be defined in terms of $\neg$ and $\rightarrow$ in the same way? Explain.

3.18 If someone says they understand what $\neg$, $\wedge$, and $\vee$ mean, but not what $\rightarrow$ means, is it possible to explain what $\rightarrow$ means from only $\neg$ and $\wedge$, or from only $\neg$ and $\vee$?

3.19 Our interpretation of $\vee$-formulas is *inclusive* in the sense that $(F \vee G)$ is true if *at least* one of the formulas F and G is true, but also if both are true. If we had wanted to capture an *exclusive* "or," we could, for example, have defined another connective, $\oplus$, and provided the following truth table:

F	G	$(F \oplus G)$
1	1	0
1	0	1
0	1	1
0	0	0

Find a way to express $(F \oplus G)$ using some of the connectives $\wedge$, $\vee$, $\neg$, and $\rightarrow$. In other words, find a formula that contains only F and G and is equivalent to $(F \oplus G)$. Justify your answer.

3.20 We may also define another connective, |, which often is called *NAND*, and interpret
(P | Q) as "not both P and Q are true simultaneously."

P	Q	(P | Q)
1	1	0
1	0	1
0	1	1
0	0	1

(a) Find a formula that is logically equivalent to ¬P in which the only connective
is |.

(b) Find a formula that is logically equivalent to (P∧Q) in which the only connective
is |.

(c) Find a formula that is logically equivalent to (P∨Q) in which the only connective
is |.

3.21 If we had defined only ⊕ and →, would it have been possible to define all the other
connectives?

3.22 Show that the following pairs of formulas are equivalent, first by logical arguments
and then using truth tables:

(a) (P → Q) and (¬Q → ¬P)

(b) (¬P ∧ ¬Q) and ¬(P ∨ Q)

(c) ¬(P ∨ ¬Q) and (¬P ∧ Q)

(d) (P ∧ ((Q ∨ R) ∨ ¬Q)) and P

(e) (P → (P ∧ ¬P)) and ¬P

(f) (P ∧ (Q ∨ R)) and ((P ∧ Q) ∨ (P ∧ R))

(g) (P → (Q → R)) and ((P ∧ Q) → R)

3.23 Three siblings, A, B, and C, live together in a house and must comply with the
following rules:

  – *If A leaves, then B must leave.*
  – *If C leaves, then if A leaves, B must stay inside.*

(a) Formalize the statements using propositional logic.

(b) Check whether it is possible for C to leave. Justify your answer.

(c) Given a language with only three propositional variables, how many inequivalent propositions can you make from these? Explain your answer well, or make
a list of all possible propositions.

(d) What is it possible for A, B, and C to do in terms of staying or leaving? List all
the possibilities.

Chapter 4

Concepts in Propositional Logic

■□■✦■■■■■■■■■■■■■■■■■■■■■

In this chapter you will learn more about semantics for propositional logic, including the concept of logical consequence. A valid argument is one whose conclusion is a logical consequence of the premises. You will also learn about the concepts of satisfiability, falsifiablility, validity, and contradiction, and how these are interrelated.

Logical Consequence

We have arrived at the fundamental question of what follows logically from what. We will soon make precise exactly what it means for a formula to follow logically from a set of assumptions, but first suppose that you know that the following formulas are true:

$$P \land Q, \quad \neg Q \lor R, \quad R \lor S.$$

Are there any other formulas that also *have* to be true in that case? The answer is yes, and another way to ask the question is this:

What follows logically from these formulas?

Try to find the answer yourself before reading further.

Because $P \land Q$ is true, both P and Q must be true, because this is how $\land$-formulas are interpreted. Because Q is true, $\neg Q$ must be false, because this is how $\neg$-formulas are interpreted. We have assumed that the formula $\neg Q \lor R$ is true, and because $\neg Q$ is false, R must be true, because this is how $\lor$-formulas are interpreted. We cannot know anything about whether S is true or false. There are also many other formulas, in fact infinitely many, that also have to be true based on our assumption, for example $\neg\neg P$, $\neg\neg Q$, $P \land R$, $Q \land R$.

The concept of *logical consequence* is important, because it accurately says what it means for something to follow logically from something else.

© Springer Nature Switzerland AG 2021
R. Antonsen, *Logical Methods*, https://doi.org/10.1007/978-3-030-63777-4_5

Definition 4.1. Logical consequence

Let S be a set of propositional formulas, and let F be a propositional formula. If F is true for all valuations that make all the formulas in S true simultaneously, then F is a **logical consequence**, or simply a consequence, of the formulas in S. We write $S \models F$ when F is a logical consequence of S.

In order to check whether F is a logical consequence of a set S, we must consider the valuations that make all the formulas in S true simultaneously and determine whether these also make F true. If that is the case, then F is a logical consequence of S. If not, then we should be able to find a valuation that makes all the formulas in S true but that makes F false.

Q **Logical consequence.** We have that Q is a logical consequence of the set that contains P and $(P \to Q)$, because all the valuations that make both P and $(P \to Q)$ true also make Q true. We may verify this in a truth table:

P	Q	P	$(P \to Q)$	Q
1	1	1	1	1
1	0	1	0	0
0	1	0	1	1
0	0	0	1	0

We can therefore write $\{P, P \to Q\} \models Q$.   ◆

❷ Is $(P \to (Q \vee R))$ a logical consequence of $(P \to Q)$?   ◆

❶ Let us check in a truth table:

P	Q	R	$(P \to Q)$	$(P$	$\to$	$(Q \vee R))$
1	1	1	1	1	1	1
1	1	0	1	1	1	1
1	0	1	0	1	1	1
1	0	0	0	1	0	0
0	1	1	1	0	1	1
0	1	0	1	0	1	1
0	0	1	1	0	1	1
0	0	0	1	0	1	0

The answer is yes, because all the valuations that make $(P \to Q)$ true, which come from all the rows except the third and fourth, also make $(P \to (Q \vee R))$ true. ◆

It is not necessary to use truth tables in order to reason about whether something is a logical consequence of something else; it is often much better to use an ordinary line of reasoning.

❷ Is $(P \to Q)$ a logical consequence of the set that contains the single propositional formula $(P \to (Q \vee R))$? ◆

❶ No, because there is a valuation that makes $(P \to (Q \vee R))$ true and $(P \to Q)$ false, namely the valuation that makes P true, Q false, and R true. This is the third row in the truth table we saw in the previous exercise. ◆

Valid Arguments

We have already seen several examples of valid arguments and valid reasoning, but now we can use the concept of *logical consequence* to define it in a precise way.

> **Definition 4.2. Valid argument**
> A line of reasoning or argument is **valid** or **sound** if the conclusion is a logical consequence of the set of premises.

Q   Two valid arguments.

If the sun is shining, then I am happy.	$(S \to H)$
The sun is shining.	S
I am happy.	H

This is an example of a *valid* or *sound* argument. If $(S \to H)$ is *true* and S is *true*, H must also be *true*. We see that H is a logical consequence of the set that consists of $(S \to H)$ and S.

If the sun is shining, then I am happy.	$(S \to H)$
I am not happy.	$\neg H$
The sun is not shining.	$\neg S$

This is also an example of a *valid* argument. A truth table can be used to analyze the argument:

S	H	$(S \to H)$	$\neg H$	$\neg S$
1	1	1	0	0
1	0	0	1	0
0	1	1	0	1
0	0	1	1	1

The bottom row is the relevant one, because this is the only place where both $(S \to H)$ and $\neg H$ are true. We see that $\neg S$ is also true here. ◆

Q **An argument that is not valid.**

If the sun is shining, I am happy.	$(S \to H)$
The sun is not shining.	$\neg S$
I am not happy.	$\neg H$

This is *not* a valid argument. This is because both $(S \to H)$ and $\neg S$ may be *true* without $\neg H$ being so. I may be happy even if the sun is not shining. In other words, $\neg H$ is *not* a *logical consequence* of $(S \to H)$ and $\neg S$. We can also see this with a single row in a truth table:

S	H	$(S \to H)$	$\neg S$	$\neg H$
0	1	1	1	0

♦

Satisfiability and Falsifiability

Definition 4.3. Satisfiability

If a valuation v makes a propositional formula F true, we say that the valuation **satisfies** the formula and write $v \models F$. A propositional formula is **satisfiable** if there is a valuation that satisfies it.

Q **Satisfiability.** The formula $(P \to Q)$ is satisfiable; it is satisfied by any valuation that makes P false or Q true. The formula $(P \land \neg P)$ is *not* satisfiable; no valuation makes both P and $\neg P$ true simultaneously.

♦

Definition 4.4. Falsifiability

If a valuation v makes a propositional formula F false, we say that the valuation **falsifies** the formula and write $v \not\models F$. A propositional formula is **falsifiable** if there is a valuation that falsifies it.

Q **Falsifiability.** The formula $(P \to Q)$ is falsifiable; it is falsified by any valuation that makes P true and Q false. The formula $(P \lor \neg P)$ is *not* falsifiable; no valuation makes both P and $\neg P$ false simultaneously.

♦

Tautology/Validity and Contradiction

> **Definition 4.5. Tautology/validity**
>
> If a propositional formula F is true for *all* valuations, we say that the formula is a **tautology**, or **valid**, and write $\models$ F.

Q Tautology. The formula $(P \vee \neg P)$ is a tautology, because any valuation will make either P or $\neg P$ true, and thus $(P \vee \neg P)$ true. The formula P is *not* a tautology, because there is a valuation that makes P false. ◆

❷ Is $((P \vee Q) \to P)$ a tautology? ◆

❶ No, because there is a valuation that makes $((P \vee Q) \to P)$ false. We can use a truth table to see this:

P	Q	((P	$\vee$	Q)	$\to$	P)
1	**1**	1	1	1	1	1
1	**0**	1	1	0	1	1
0	**1**	0	1	1	0	0
0	**0**	0	0	0	1	0

The relevant valuation may be read from the third row. It makes P false and Q true. ◆

❷ Is $(P \to (P \vee Q))$ a tautology? ◆

❶ Yes, because any valuation makes $(P \to (P \vee Q))$ true. The reason for this is that if a valuation makes P true, it will also make $(P \vee Q)$ true. ◆

❷ Is $(P \to P)$ a tautology? ◆

❶ Yes, because any valuation makes $(P \to P)$ true. ◆

> **Definition 4.6. Contradiction**
>
> If a propositional formula F is false for *all* valuations, we say that the formula is **contradictory**, or a **contradiction**.

Q Contradictory. The formula $(P \wedge \neg P)$ is contradictory because no valuation makes both P and $\neg P$ true simultaneously; any valuation will either make P false or $\neg P$ false, and thus $(P \wedge \neg P)$ false. The formula P is *not* contradictory, because there is a valuation that makes P true. ◆

❷ Is ¬P a contradiction?   ◆

❶ No, because there is a valuation that makes ¬P true.   ◆

❷ Is ¬(P → P) a contradiction?   ◆

❶ Yes, because any valuation makes ¬(P → P) false. We can see this using a truth table:

P	¬	(P → P)
1	0	1
0	0	1

◆

❷ Is (P → P) a contradiction?   ◆

❶ No, because there is a valuation that makes (P → P) true.   ◆

Here is a table summarizing all these terms:

	F *is true*	F *is false*
for all valuations	F *is valid*	F *is contradictory*
for at least one valuation	F *is satisfiable*	F *is falsifiable*

Symbols for Truth Values

It is often useful to have symbolic representations for the truth values **1** and **0**. It is possible to use (P ∨ ¬P), which is always true, and (P ∧ ¬P), which is always false, for this, but that is a little cumbersome.

Definition 4.7. Symbols for the truth values

We consider ⊤ and ⊥ to be propositional formulas in addition to the ones we already have. It is common to read ⊤ as "top" or "true" and ⊥ as "bottom" or "false." Every valuation must make ⊤ true and ⊥ false.

We get that ⊤ and ⊥ combine with the other connectives in the following way:

$$⊤ → A ⇔ A \quad \text{and} \quad A → ⊥ ⇔ ¬A,$$
$$⊤ ∧ A ⇔ A \quad \text{and} \quad A ∧ ⊥ ⇔ ⊥,$$
$$⊤ ∨ A ⇔ ⊤ \quad \text{and} \quad A ∨ ⊥ ⇔ A.$$

Connections Between Concepts

That a formula is *valid* can be used to express a variety of other properties, for example both *logical equivalence* and *logical consequence*. That two formulas F and G are *logically equivalent* means that they always have the same truth value, no matter what valuation we have. But this is exactly the same as saying that the formula F ↔ G is *valid*. These are two ways of expressing exactly the same thing:

$$F \Leftrightarrow G \quad \text{if and only if} \quad \models F \leftrightarrow G.$$

Nevertheless, be aware of the difference between these statements: On the left side it says F ⇔ G. This is a statement about two propositional formulas, which says that they are equivalent. On the right side it says ⊨ F ↔ G. This is a statement about *one* propositional formula, which says that it is valid.

We can do exactly the same for the property *logical consequence*. That G is a *logical consequence* of the set consisting of F means that every valuation making F true also makes G true. But this is exactly the same as saying that the formula F → G is *valid*. We may therefore write the following:

$$\{F\} \models G \quad \text{if and only if} \quad \models F \to G.$$

If we let F ⇒ G mean that G is a logical consequence of the set consisting of F, we may express this connection like this:

$$F \Rightarrow G \quad \text{if and only if} \quad \models F \to G.$$

We also have that two formulas F and G are logically equivalent exactly when they are logical consequences of each other. With the notation above, we can express this in the following way:

$$F \Leftrightarrow G \quad \text{if and only if} \quad F \Rightarrow G \text{ and } G \Rightarrow F.$$

Notice that only → and ↔ belong to the *object language*, the formal propositional language. The other symbols, ⇒, ⇔, and ⊨, belong to the *metalanguage* and are *abbreviations* for statements that we might as well write with words.

Independence of Formulas

The concept of *logical consequence* can be used to define exactly what it means for formulas to be *independent* of each other.

Definition 4.8. Independent formula and set of formulas

A formula F is **independent** of a set of formulas S if neither F nor ¬F is a logical consequence of S. A set of formulas is **independent** if each formula in the set is independent of the set consisting of all the other formulas in the set.

Q **Independent formulas and sets.** The formula P is independent of the set $\{P \vee Q, R\}$ because neither P nor ¬P is a logical consequence of it. But P is not independent of the set $\{P \wedge Q, R\}$, because P is a logical consequence of it. The set $\{P \wedge Q, P \vee Q\}$ is *not* independent, because $P \vee Q$ is a logical consequence of $\{P \wedge Q\}$. The set $\{P, P \rightarrow Q\}$ is independent, because neither P nor ¬P follows from $\{P \rightarrow Q\}$, and neither $P \rightarrow Q$ nor ¬$(P \rightarrow Q)$ follows from $\{P\}$. The set $\{P, P \rightarrow Q, Q\}$ is *not* independent, because Q is a logical consequence of $\{P, P \rightarrow Q\}$.   ♦

Deciding Whether a Formula Is Valid or Satisfiable

It is possible to write a computer program that can check whether a propositional formula is valid and that always gives the correct answer in a finite amount of time. An easy way to do this is by setting up a truth table and checking each row for what the truth value of the formula is, but that is very inefficient. A better method is to assume that the formula is not valid and show that this is impossible. Checking whether a formula is valid can be translated into checking whether a formula is satisfiable. In order to check whether a formula F is valid, we can check whether the formula ¬F is satisfiable. If it is, F is not valid; if it is not, F is valid. Determining whether a formula is satisfiable is referred to as the **satisfiability problem**. There are many ways to check satisfiability, but nobody knows what the limit is for how efficiently it is possible to do this. This is one of the major unsolved problems in mathematics and computer science, called the **P = NP** problem.

Digression

The **P = NP** problem is about the efficiency of algorithms and how long it takes for an algorithm to finish depending on the size of the input. Some algorithms use polynomial time, which means that the time it takes before the algorithm finishes may be expressed as a polynomial in the size of the input, for example a square or an nth power. Such algorithms are considered to be efficient, even though they can use a very long time. **P** stands for the set of problems that can be solved by algorithms in polynomial time. **NP** stands for the set of problems that can be solved by algorithms in *nondeterministic* polynomial time, which means that the algorithm can spend unlimited (greater than polynomial) time finding the solution, but that *checking whether the solution is correct can be done in polynomial time*. The big question is whether **P** and **NP** are the same, that is, whether there are efficient algorithms for the problems in **NP**. The "hardest" problems in **NP** are called **NP-complete**. Here are some examples of these: To decide whether a propositional formula is satisfiable. To decide whether a graph is isomorphic to a part of another graph. To decide whether there is a Hamiltonian path or cycle in a graph. There are also problems related to games like Minesweeper, Tetris, and Sudoku that are **NP**-complete. To solve the **P = NP** problem, it is sufficient to find an efficient algorithm for any of the **NP**-complete problems, or prove that no such thing can exist, but nobody has yet managed to do either of these.

Exercises

4.1 Is $(P \lor Q)$ a logical consequence of $(P \land Q)$? Justify your answer.

4.2 Decide whether each of the following arguments is valid. Justify your answer.

Argument (a):		Argument (b):	
Premises:	$(A \lor B)$	*Premises:*	
	$(A \to C)$		$(A \land B) \to C$
	$(B \to C)$		A
Conclusion:	C	*Conclusion:*	C

4.3 Is $P \to (Q \to P)$ a tautology? Justify your answer.

4.4 Show that the following formulas are valid by setting up truth tables:

(a) $(P \lor (Q \lor \neg P))$ (b) $(P \land (P \to Q)) \to Q$ (c) $(P \to (P \lor Q))$

4.5 Show that the following formulas may be falsified:

(a) $(P \land \neg Q) \lor (\neg P \land Q)$ (b) $(P \to Q) \lor (R \to Q)$ (c) $P \to (P \land Q)$

4.6 Show that the following formulas are contradictions by setting up truth tables:

(a) $(P \land \neg P)$ (b) $(P \lor Q) \land (\neg P \land \neg Q)$ (c) $\neg(P \to Q) \land Q$

4.7 Show that the following formulas are satisfiable:

(a) $(P \lor Q) \land (\neg P \lor \neg Q)$ (c) $(P \to (Q \to R)) \land (Q \land \neg R)$
(b) $(P \to Q) \land \neg Q$

4.8 Which of the formulas below are tautologies, and which contradictions? Justify your answers. If any of the formulas below are neither tautologies nor contradictions, explain why that is the case.

(a) $(P \lor Q) \to \neg P$ (d) $((P \to Q) \land \neg Q) \to \neg P$
(b) $P \lor (Q \to \neg P)$ (e) $\neg(P \lor Q) \land (\neg Q \lor R) \land (\neg R \lor P)$
(c) $(P \land Q) \to \neg P$ (f) $\neg(P \lor Q) \land P$

4.9 Each formula below is either a tautology or a contradiction. Decide which formulas are which. Explain using truth tables or give a semantic argument.

(a) $(\neg B \lor (C \to (A \land D))) \lor (B \to C)$ (d) $B \land \neg((C \land A) \to B)$
(b) $((B \to C) \leftrightarrow D) \to ((A \land D) \to D)$ (e) $\neg(\neg C \to ((D \land B) \to D))$
(c) $((C \land (D \lor A)) \to B) \lor \neg B$ (f) $(A \to B) \land (B \to C) \to (\neg C \to \neg A)$

4.10   Find propositional formulas F, G, H, and I such that all of the following statements are true. Hint: You need only two propositional variables.

- I is a logical consequence of G.
- G is a logical consequence of F.
- I is a logical consequence of H.
- H is a logical consequence of F.
- G is *not* a logical consequence of H.
- H is *not* a logical consequence of G.

4.11   Here are some propositional formulas. Place ⇒-arrows that indicate which formulas are logical consequences of which. That is, place an arrow from a formula F to a formula G if G is a logical consequence of F. It is not necessary to place an arrow from a formula to itself.

$$(P \wedge Q) \qquad (P \vee Q)$$

$$(P \rightarrow Q) \qquad (P \wedge \neg P)$$

4.12   A set of formulas is **satisfiable** if there is a valuation that makes all the formulas of the set true simultaneously. Suppose that a set S of formulas is satisfiable and that a formula F is independent of S. Prove that both $M \cup \{F\}$, the set we get by adding F, and $M \cup \{\neg F\}$, the set we get by adding ¬F, are satisfiable.

4.13   Is it the case that "the formula G is a logical consequence of the formula F" and "the formula F → G is a tautology" mean exactly the same thing?

4.14   For each of the claims below, determine whether it is true or false. Justify your answer.

(a) The formula Q is independent of the set $\{P \vee Q, R\}$.
(b) The formula Q is independent of the set $\{P \wedge Q, R\}$.
(c) The formula Q is independent of the set $\{\neg Q, P \vee R\}$.
(d) The set $\{P, Q, R\}$ is independent.
(e) The set $\{P \wedge Q, P \vee Q\}$ is independent.
(f) The set $\{P, P \rightarrow Q, \neg Q\}$ is independent.

4.15   Is it true that every propositional formula is a logical consequence of ⊥? Explain why or why not. If it is true, provide a proof that such is the case; if it is not true, provide a counterexample, that is, a propositional formula that is not a logical consequence of ⊥.

Chapter 5
Proofs, Conjectures, and Counterexamples

■□□■▌■■■■■■■■■■■■■■■■■■■□

In this chapter you will learn some methods of proof and become familiar with the relationships between proofs, conjectures, and counterexamples. The goal is to become better at proving statements. Several common methods of proof are discussed: direct proofs, existence proofs, proofs for universal statements, contrapositive proofs, proofs by contradiction, proofs that something is not true, proofs by cases, as well as the difference between constructive and nonconstructive proofs.

Proofs

Science is about finding out what is true, both concretely and abstractly, and when we have a *proof* of a statement, we have a guarantee that the statement is true. The quest for proof is about obtaining absolute certainty. *Without proof, how can we be sure that something is true?* This is one of the reasons we are so interested in proofs.

Definition 5.1. Proof

A **proof** that a statement follows from a given set of assumptions is a series of logical conclusions that shows how to get from the assumptions to the statement. For each step, the conclusion must be a *logical consequence* of the assumptions.

What is possible to prove depends directly on what we assume. The question of what is true quickly boils down to determining what follows logically from a set of assumptions.

You can assume anything, but you must then also accept the consequences. It is a bit like with cooking or music: When we cook, there is no limit to what ingredients we can use and how we combine them, but not everything will taste good. And if we sit at the piano, there are many possibilities for what we can play, but not everything will sound good. That is the case for mathematics, too: you are allowed to assume anything, but not everything will lead somewhere or be useful. Now we will look at some of the most common methods for proving that something is true. In Chapters 11 and 12, we will look at two additional methods.

© Springer Nature Switzerland AG 2021
R. Antonsen, *Logical Methods*, https://doi.org/10.1007/978-3-030-63777-4_6

Conjectures

We call statements that we think are likely true but for which we have not found proofs *conjectures*. Mathematics is full of such conjectures and unsolved problems. Some are very complex and difficult to understand, but there are also many that are easy to formulate.

Definition 5.2. Conjecture

A **conjecture** is a statement that we believe, or have good reason to believe, is true, but that we have neither proved nor disproved.

Goldbach's conjecture states that every even number greater than two may be expressed as the sum of two prime numbers. Try it for yourself! The conjecture is named after the German mathematician *Christian Goldbach* (1690–1764), who formulated it in 1742 in a letter to *Leonhard Euler* (1707–1783). For example:

$$4 = 2+2, \qquad 8 = 3+5, \qquad 12 = 5+7, \qquad 16 = 3+13,$$
$$6 = 3+3, \qquad 10 = 3+7, \qquad 14 = 3+11, \qquad 18 = 5+13.$$

Nobody knows whether this conjecture is true; no one has managed to prove or disprove the statement. One way to disprove the statement would be to find a **counterexample**, an even number greater than 2 that *cannot* be written as the sum of two prime numbers. We have, of course, checked very many numbers.

The **Collatz conjecture** from 1937 is another famous conjecture, named after the German mathematician *Lothar Collatz* (1910–1990). Choose a natural number n. If n is even, divide it by 2; if n is odd, multiply it by 3 and add 1. Then repeat this process over and over, stopping if you reach the number 1.

For example, if we start with the number 6, we get 6, 3, 10, 5, 16, 8, 4, 2, and finally 1. The Collatz conjecture says that this process will always end, no matter what number we start with.

There is no known solution to this problem. We have found neither a *proof* that every sequence will end with 1 nor a *counterexample* in the form of an infinite sequence that never ends. All numbers up to over five billion billion have been checked, and the sequence always ends with 1, but we are still not sure whether this happens with every number.

THE COLLATZ CONJECTURE STATES THAT IF YOU PICK A NUMBER, AND IF IT'S EVEN DIVIDE IT BY TWO AND IF IT'S ODD MULTIPLY IT BY THREE AND ADD ONE, AND YOU REPEAT THIS PROCEDURE LONG ENOUGH, EVENTUALLY YOUR FRIENDS WILL STOP CALLING TO SEE IF YOU WANT TO HANG OUT.

(Source: xkcd.com/710)

The procedure in the Collatz conjecture is a good example of how a *simple* rule may give rise to something of enormous complexity.

Pólya's conjecture is another good example. Every natural number greater than 1 can be written uniquely as a product of prime numbers, and we can count the number of factors as follows. Let us call those numbers with an odd number of factors ODD and the others EVEN.

$2 = 2$ has *one* factor. ODD	$6 = 2 \cdot 3$ has *two* factors. EVEN
$3 = 3$ has *one* factor. ODD	$7 = 7$ has *one* factor. ODD
$4 = 2 \cdot 2$ has *two* factors. EVEN	$8 = 2 \cdot 2 \cdot 2$ has *three* factors. ODD
$5 = 5$ has *one* factor. ODD	$9 = 3 \cdot 3$ has *two* factors. EVEN

ODD	2	3		5		7	8			11	12	13					17	18
EVEN					4		6		9	10				14	15	16		

George Pólya (1887–1985) noticed the following: it looks as though there are always more ODD numbers than EVEN numbers as we go up the number line. For example, when we have arrived at 18, we have encountered seven EVEN and ten ODD. *Pólya's conjecture* from 1919 was that there are always more ODD than EVEN, no matter how far we go.

The conjecture holds for the first few million numbers. It actually holds for *ten million* numbers. And for a *hundred million* numbers. But it does not hold! At 906 150 257 ODD is no longer ahead. A proof was found in 1958, but it was not *constructive*. There was no concrete counterexample, but only an estimate that there had to be a counterexample around 10^{361}. In 1960, the counterexample 906 180 359 was found, and in 1980, the smallest counterexample, which is 906 150 257. This is one of the many reasons we want *proofs* for our statements.

Thinking from Assumptions

Logic is the art of thinking from assumptions. We can assume anything; the interesting thing is what *logically follows* from those assumptions.

Q **Logical consequence.** Assume (1) that $((P \to Q) \wedge P)$ is true. From (1) it follows (2) that $(P \to Q)$ is true, and (3) that P is true. From (2) and (3) it follows (4) that Q is true. From (3) and (4) it follows that $(P \wedge Q)$ is true. From the *assumption* that $((P \to Q) \wedge P)$ is true, it therefore follows that $(P \wedge Q)$ is true. In other words, $(P \wedge Q)$ is a *logical consequence* of $((P \to Q) \wedge P)$. ◆

Q **Logical consequence.** Assume (1) that *if* the king arrives, *then* it is party time. *What follows logically from this assumption?* Does it follow that the king arrives? No. Does it follow that it is party time? No. Assume additionally (2) that both the king and the prime minister arrive. Now we have *two* assumptions. *What follows from these two assumptions?* From (2) it follows (3) that the king arrives. From (1) and (3) it follows that it is party time. Instead of (2), assume now (4) that it is *not* party time. From (1) and (4) it follows that the king is not coming. Why? Because (1) and (4) cannot be true at the same time as the king arrives. ◆

Direct Proofs

Definition 5.3. Direct proof

A **direct proof** of a statement of the form "if F, then G" is a logically valid argument that begins with the assumption that F is true and ends with the conclusion that G is true.

A direct proof typically begins with "suppose that F is true" and ends with "it therefore follows that G is true." The following examples are a bit artificial, but they show the *structure* of direct proofs.

❷ Suppose that a valuation is given. Prove the statement "if the valuation makes $(P \land Q)$ true, then the valuation makes P true."   ◆

❶ Suppose that the valuation makes $(P \land Q)$ true. Then the valuation must make both P and Q true, because that is how $\land$-formulas are interpreted, and in particular the valuation must make P true. It therefore follows that if $(P \land Q)$ is true, then P is true.   ◆

❷ Suppose that we have a valuation that makes $(P \rightarrow Q)$ true. Prove the statement "if P is true, then $(Q \lor R)$ is true."   ◆

❶ We are given (1) that a certain valuation makes $(P \rightarrow Q)$ true. To prove the statement, we assume (2) that the valuation makes P true. Our goal is now, from this assumption, to show that the valuation makes $(Q \lor R)$ true. From (1) and (2) it follows that Q is true, because of how $\rightarrow$-formulas are interpreted. Then $(Q \lor R)$ must also be true, because of how $\lor$-formulas are interpreted. We see that the assumption that P is true leads to the conclusion that $(Q \lor R)$ is true. We can therefore conclude that the valuation makes $(P \rightarrow (Q \lor R))$ true.   ◆

Note that the proofs in these examples do not contain any instances of the words "not" or "false" or the connective $\neg$. That is because such negatives are not needed. It is a good exercise to try to prove statements with as few references to falsehood as possible. *If it is not necessary to refer to negation or falsehood, why should we do it?* This is an exercise in thinking from assumptions. If we are to prove that **Y** is a logical consequence of **X**, we can assume that **X** is *true* and from *this assumption* alone show that **Y** is also true. A direct proof of a statement is also often easier to understand in retrospect than one requiring negative assumptions.

❷ Prove the statement "if P is false, then $P \lor (P \land Q)$ is false."   ◆

❶ Suppose that P is false. Then $P \land Q$ must also be false, because of how $\land$-formulas are interpreted. Then, $P \lor (P \land Q)$ must also be false, because of how $\lor$-formulas are interpreted. Thus we have shown that if P is false, $P \lor (P \land Q)$ is false.   ◆

Here is an example involving even numbers. We say that an integer x is an **even number** if there is an integer y such that $x = 2y$. This definition is used in several places in the following proof.

❷ Let x be an integer. Show that if x is an even number, then x^2 is an even number.◆

❶ Suppose that x is an even number. Then x equals $2y$ for some integer y, by the definition of even numbers. Then x^2 equals $(2y)^2$, or $4y^2$. To show that x^2 is an even number, it is sufficient to show that it is equal $2z$ for some integer z. We may write $4y^2$ as $2(2y^2)$, and $2y^2$ must be an integer, because y is an integer. Setting $z = 2y^2$, we have that $x^2 = 2z$, and it follows that x^2 is an even number. We have thus proved the statement.   ◆

When we want to show that a statement of the form "if F, then G" is true, we begin by assuming that F is true, and from this we show that G also must be true. *Why do we not also have to look at the case in which F is false?* Because we don't need to. There is only one case in which a statement of the form "if F, then G" becomes false, and that is when F is true and G is false. A direct proof *excludes* this possibility. There is no need to check what follows from the assumption that F is false. In fact, if we assume that F is false, we indirectly assume that "if F, then G" is true, which is exactly what we are showing.

Existence Proofs

An **existential statement** is a statement that says that something exists. How do we prove existential statements? We prove existential statements simply by finding an object that makes the statement true. Such proofs are called **existence proofs**.

❷ Prove the statement "there is a natural number x such that $x + x = 8$."   ◆

❶ Let x be the number 4. Then $x + x$ must equal $4 + 4$, which equals 8.   ◆

❷ Prove the statement, "there is a valuation that makes the formula $P \rightarrow \neg P$ true." ◆

❶ Choose a valuation that makes P false. This makes $P \rightarrow \neg P$ true.   ◆

Proofs by Cases

Another common method of proof is **proof by cases** or **proof by exhaustion**. Here, a proof is divided into smaller parts, which together cover what we need to prove, and each part is proved individually.

❷ Let $A = \{1, 2\}$ and $B = \{1, 2, 3, 4\}$. Show that $A \subseteq B$.   ◆

❶ Let x be an element of A. We must from this assumption show that x is an element of B. Either $x = 1$ or $x = 2$. In both cases, x is an element of B. We can conclude that x is an element of B, and therefore that $A \subseteq B$.   ◆

Proofs of Universal Statements

Universal statements are statements that say something about all objects of a particular type. For example "all even numbers are divisible by 2" or "any valuation that makes P true must make $P \lor Q$ true." How do we prove universal statements? A common way is to choose an **arbitrary** object from a given set and show that *this object* has the desired property. The word "arbitrary" means that there are no additional constraints on which object we choose; it might as well have been any other object in the set. In other words, we should not assume anything additional about the object. If we want to be precise, in the end we can say "because the object was arbitrarily chosen, the statement holds for all objects." The following examples are a bit artificial, but they show in a nice way the *structure* of a proof of a universal statement.

❓ Prove the following statement: "all even numbers are divisible by 2."   ◆

❗ This is a universal statement because it says something about *all* even numbers. Let x stand for an arbitrary even number. By definition, x must be of the form 2y, where y is an integer. Then the number can be divided by 2. *Because x was an arbitrarily chosen even number, it follows that all even numbers are divisible by 2.*   ◆

❓ Prove the statement, "a valuation that makes P true also makes $(P \lor Q)$ true."   ◆

❗ This is a universal statement because it says something about *all* valuations. Choose an arbitrary valuation, and suppose that it makes P true. By the definition of truth tables for $\lor$-formulas, it must make $(P \lor Q)$ true. Because the valuation was arbitrarily chosen, it follows that any valuation that makes P true must make $(P \lor Q)$ true.   ◆

Why does this always work? The reason is that when we choose an arbitrary element from a set, it can represent any element in the set. The element becomes a *placeholder* for any of the other elements. If the argument works for this element, it must also work for all the others. Therefore, the argument must hold for *all* elements.

It is common to distinguish between the words *random* and *arbitrary*. If we choose an item randomly, it means that the probability of choosing it is equal to that of choosing any other element. An arbitrary element, on the other hand, is a placeholder that acts as a representative of any of the other elements. When we talk about arbitrary elements, we always use a variable or a placeholder. The number 7 may, for example, be a randomly chosen number less than one hundred, but it cannot be an arbitrary number. In order to prove that a statement is true for all numbers, it is not sufficient to choose a random number; the number must be *arbitrary*.

❓ Show that Q is a logical consequence of $(\neg P \land \neg\neg Q)$.   ◆

  – First attempt at a solution: *Suppose that all valuations make $(\neg P \land \neg\neg Q)$ true.* This is not a good idea, because what we are supposing is, in fact, false. From this assumption all sorts of things follow, among them that Q is true, but it does not contribute to solving what the exercise asks for.

– Second attempt at a solution: *Suppose that* $\neg\neg Q$ *is true.* This is also not a good idea. It does, of course, provide that Q is true, but this means only that Q is a logical consequence of $\neg\neg Q$, and that is not what the exercise asks for.

❶ *Suppose that* $(\neg P \wedge \neg\neg Q)$ *is true.* From this assumption it follows that $\neg\neg Q$ is true, and thus that Q is true. This shows that Q is a logical consequence of $(\neg P \wedge \neg\neg Q)$.♦

Counterexamples

If a statement is not true, it is impossible to prove it. If the statement is *universal*, it is in principle possible to find a **counterexample** to it. A counterexample is a form of existence proof; it is a proof that *there exists* a case that makes the statement false.

Q Counterexample. The statement "P is a logical consequence of $(P \vee Q)$" is not true. The valuation that makes P false and Q true is a counterexample. It makes $(P \vee Q)$ true, but P false, and therefore shows that P is not a logical consequence of $(P \vee Q)$.♦

❷ Find a counterexample to the statement, "if G is a logical consequence of F, then F and G are equivalent."   ♦

❶ If we try to prove this statement, we quickly find that we are getting nowhere. When we cannot find a proof, it often pays to start looking for a counterexample. In this case, a counterexample can be two concrete formulas F and G such that G is a logical consequence of F, but the formulas are *not* equivalent. One possibility for this is that F is P and that G is $(P \vee Q)$. Then the "if"-part is true, because $(P \vee Q)$ is a logical consequence of P, but the "then"-part is false, because $(P \vee Q)$ and P are not logically equivalent.   ♦

Contrapositive Proofs

Sometimes it is difficult to complete a direct proof of a statement of the form "if F, then G." Then it is possible to start with the assumption that G is false and end with the conclusion that F is false. Then the original statement must hold, and this is called proving the contrapositive.

> **Definition 5.4. Contrapositive proof**
>
> A **contrapositive proof** of a statement of the form "if F, then G," is a logically valid argument that begins with the assumption that G is false and concludes that F is false. The **contrapositive** of the formula $(F \to G)$ is the formula $(\neg G \to \neg F)$.

An $\to$-formula and its contrapositive formula are always equivalent. Also notice that a contrapositive proof of $(F \to G)$ is a direct proof of the contrapositive, $(\neg G \to \neg F)$.

It is good practice to say that we have proved something contrapositively if that is what we have done. Sometimes it is easier to prove a statement contrapositively, but a direct proof is usually to be preferred. A direct proof can be much easier to read, understand, and convey.

❷   Prove the statement "if $3n + 2$ is an odd number, then n is an odd number."   ◆

❶   We prove the contrapositive statement, which is equivalent: "if n is not an odd number, then $3n + 2$ is not an odd number." From here, the structure of the proof is identical to a direct proof. Suppose that n is not an odd number, that is, that n is an even number. Then n is of the form $2x$ for some integer x. Then $3n + 2$ equals $3(2x) + 2$, which equals $6x + 2$. Because this number equals $2(3x + 1)$, it is an even number. Therefore, $3n + 2$ is not an odd number. We have thus proved the original statement by proving its contrapositive.   ◆

Proofs by Contradiction

> **Definition 5.5. Proof By Contradiction**
>
> A **proof by contradiction** of a statement is a proof that begins with the assumption that the statement is false and shows how this assumption leads to a contradiction. The proof concludes that the statement must be true.

If we assume that a statement is not true and this leads to a contradiction, we conclude that the statement must be true. Note that the conclusion is positive; it says that the statement is true. In a proof by contradiction, we assume the opposite of what we want prove, and if this leads to a contradiction, the assumption must have been false, and we have therefore proved what we wanted.

❷   Prove that the formula $(P \to Q) \lor (Q \to P)$ is true for all valuations.   ◆

❶   Suppose for the sake of reaching a contradiction that the statement does not hold. Then there must be a valuation that makes the formula false. The only way to make a disjunction false is by making both of the disjuncts false. Then this valuation must make both $(P \to Q)$ and $(Q \to P)$ false. Since the valuation makes $(P \to Q)$ false, it must make P true and Q false. Since the valuation makes $(Q \to P)$ false, it must make Q true and P false. It is not possible for a valuation to make P both true and false, and we have a contradiction. We conclude that $(P \to Q) \lor (Q \to P)$ is true for all valuations.   ◆

In a proof by contradiction we show that something is true by showing that it is impossible that it is false. Such an argument, however, requires the assumption that a statement is true if and only if it is not false, but this is the case for classical propositional logic. Proofs like these usually begin with "suppose for the sake of a contradiction that the statement does not hold." This principle is called **reductio ad absurdum** – a reduction of something to the absurd. It may sometimes be easier to prove something with a proof by contradiction, but most often it is more than we need. Proofs by contradiction are also not *constructive* in the same sense as direct proofs or proofs in which objects are created.

Constructive Versus Nonconstructive Proofs

A proof by contradiction is an example of a *nonconstructive* proof. A proof by contradiction of a statement that something exists does not need to provide an object that makes the statement true. That is what makes it a **nonconstructive proof**. If the proof displays or provides a method for finding the object, it is **constructive**. Suppose for example that we want to prove the statement "there is life on Mars." A proof by contradiction would begin with the assumption that there is *not* life on Mars, and end with a contradiction, in order to conclude that there actually is life on Mars. Many believe that such a proof is not good enough. A constructive proof is one in which we actually find life on Mars and exhibit it.

Digression

The mathematical strategy game **Chomp** was invented by the American mathematician *David Gale* (1921–2008). The game is based on a rectangular chocolate bar. Two players take turns choosing a square and eating the chosen square and all the squares that are located above and to the right of it. The square at the bottom left (■) is poisoned, and the player who eats it loses. Here are two different games with a chocolate bar with three rows and four columns. We see that player 1 (■) defeats player 2 (■) in both games:

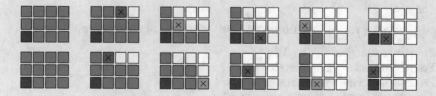

We can prove that there is a strategy that ensures that player 1 *always* wins, regardless of the size of the chocolate bar: Suppose that player 2 has a winning strategy, no matter how player 1 begins. Let player 1 start by selecting the top right-hand square. By assumption, player 2 has a strategy, and therefore a move, that is winning. But player 1 could just as well have started with that move. In other words, if player 2 had a winning strategy, player 1 could just have *copied* that strategy and won the game. This proves that player 1 has a winning strategy, no matter how many rows and columns in the chocolate bar. But the proof is *nonconstructive* and does not provide a concrete strategy that we can actually use to win! Although we know that player 1 always has a winning strategy, the proof does not provide any information about what it is.

Proofs of Falsity

If we want to prove that a statement is *not* true, we can start by assuming that the statement is true and show how that leads to a contradiction. This is *not* the same as a proof by contradiction. It is a direct proof that something is false. This is a subtle, but important, difference, which we can illustrate with propositional logic: A proof by contradiction that a statement P is true is a proof that ¬P leads to a contradiction. A proof that a statement P is not true is a proof that the assumption that P is true leads to a contradiction.

The left-hand figure illustrates a *proof of* P *by contra-diction*. Here we see the *indirect* structure of such a proof, because we start with the opposite of what we want to prove.

The right-hand figure illustrates a *direct proof of* ¬P. Here we see the *direct* structure of such a proof, because we start with an assumption and show that this *cannot* be true.

$$
\begin{array}{cc}
\neg P & P \\
\vdots & \vdots \\
\bot & \bot \\
\hline
P & \neg P
\end{array}
$$

We will meet this difference again in Chapter 24 when we look at *natural deduction*, a logical calculus to formalize derivations and proofs.

❷ Show that formulas P and (P → ¬P) cannot be true simultaneously.   ◆

❶ Suppose that P and (P → ¬P) are true simultaneously. Then there is a valuation that makes P true and (P → ¬P) true. By the interpretation of →-formulas, the valuation must also make ¬P true. By the interpretation of ¬-formulas it follows that P is false. But that is impossible, because we have assumed that the valuation makes P true. We can conclude that the formulas cannot be true simultaneously.   ◆

Finally, here are some "methods of proof" that you should possibly avoid:

- Proof by *persuasion*: "It's trivial!"
- Proof by *omission*: "The rest is left to the reader."
- Proof by *authority*: "The lecturer said so."
- Proof by *consensus*: "Everyone knows it."
- Proof by *picture*: "Look here!"
- Proof by *cosmology*: "The negation is impossible to imagine."
- Proof by *parents*: "That's just how it is."
- Proof by *hand-waving*: "And you just continue like this."

Exercises

5.1 Here is a series of statements. Some are true and some are false. Discuss what it would take to be able to *disprove* each of the statements. You may answer this in general without knowing exactly what all the terms in the statements mean.

(a) *If the ball is red, then it is not green.*

(b) *The number x is divisible by 2 or 3.*

(c) *There are infinitely many perfect numbers.*

(d) *The number x is divisible by both 2 and 3.*

(e) *There is an even number greater than 2 that cannot be written as the sum of two prime numbers.*

(f) *All students who do the exercises will pass the exam.*

(g) *It is not the case that all even numbers are divisible by 4.*

5.2 Prove the following statements by contradiction.

(a) *If $3n + 2$ is an odd number, then n is an odd number.*

(b) *If $(P \vee Q)$ is true and $\neg P$ is true, then is Q true.*

(c) *All valuations make $P \rightarrow (Q \rightarrow P)$ true.*

(d) *If $A \subseteq (B \cap C)$, then $A \subseteq B$.*

5.3 What is the connection between a contrapositive proof and a proof by contradiction? Explain by means of propositional logic.

5.4 Explain what a counterexample would be for the following statements:

(a) *All politicians are wise.*

(b) *If you come to the party, then I will be happy.*

(c) *You do the exercises or fail the exam.*

(d) *The book was exciting, and I forgot about time.*

(e) *All prime numbers are odd numbers.*

(f) *All integers that are divisible by 3 are also divisible by 6.*

(g) *All integers that are divisible by both 4 and 6 are also divisible by 24.*

(h) *For all natural numbers n, n^3 is greater than $2^n - 1$.*

(i) *If x is an element of A, then x is an element of $A \cap B$.*

(j) *If $A \subseteq B$, then $A \subseteq (B \cap C)$.*

(k) *All valuations make P true or $(\neg P \wedge Q)$ true.*

(l) *If $(P \rightarrow Q)$ is true, then Q is true.*

5.5 Prove the statement "If $(P \rightarrow Q)$ and $(Q \rightarrow R)$ are true, then $(P \rightarrow R)$ is true" in the following ways: (a) With a direct proof. (b) With a contrapositive proof. (c) With a proof by contradiction.

5.6   Look at the formula $P_1 \wedge P_2 \wedge P_3 \to Q_1 \vee Q_2 \vee Q_3$. What is needed in order to show that such a formula is false?

5.7   Can you conclude that $A = B$ if A, B, and C are sets such that the following statements hold? If that is the case, provide a proof; if it is not the case, provide a counterexample.

$$A \cup C = B \cup C, \qquad A \cap C = B \cap C.$$

5.8   For each of the statements below, decide whether it is true or false. We assume that F and G represent propositional formulas. If the statement is true, provide a proof or explain why it is the case. If the statement is false, provide a counterexample:

(a) *If F is valid and G is valid, then $(F \wedge G)$ is valid.*

(b) *If $(F \wedge G)$ is valid, then F is valid and G is valid.*

(c) *If F is satisfiable and G is satisfiable, then $(F \wedge G)$ is satisfiable.*

(d) *If $(F \wedge \neg G)$ is a contradiction, then $(F \to G)$ is valid.*

(e) *If F is satisfiable, then F is falsifiable.*

(f) *If G is satisfiable, then $(F \to G)$ is satisfiable.*

5.9   For each of the statements below, decide whether it is true or false. We assume that F, G, and H represent propositional formulas. If the statement is true, provide a proof or explain why it is the case. If the statement is false, provide a counterexample.

(a) *If $\neg F$ is satisfiable, then F is valid.*

(b) *If F is not satisfiable, then $\neg F$ is satisfiable.*

(c) *If F is satisfiable, then $\neg F$ is not satisfiable.*

(d) *If F is not a logical consequence of G, then $\neg F$ must be a logical consequence of G.*

(e) *If $F \to G$ is valid, then $F \to (G \vee H)$ is valid.*

(f) *If $F \to G$ is valid, then $F \to (G \wedge H)$ is valid.*

(g) *For all F, F is satisfiable or $\neg F$ is satisfiable.*

(h) *For all F, F is valid or $\neg F$ is valid.*

(i) *If logic is hard and fun, then logic is fun.*

5.10   Decide whether the following arguments are valid. If the argument is valid, explain why; if not, provide a counterexample that shows that it is not valid.

(a) *If x is a real number such that $x > 1$, then $x^2 > 1$. Suppose that $x^2 > 1$. Then it follows that $x > 1$.*

(b) *The formula G is a logical consequence of the formula F. Suppose that F is false. Then G must be false.*

5.11 Express the following sentences using propositional logic. Represent the three sentences as propositional formulas, and investigate whether it is possible for the formulas to be true simultaneously. What does this say about the limitations of propositional logic?

(a) *If Ola lives in Paris, then he lives in France.*

(b) *It is not true that if Ola lives in London, then he lives in France.*

(c) *It is not true that if Ola lives in Paris, then he lives in England.*

5.12 For each of the following statements, determine whether it is true or false? If the statement is false, provide a counterexample. If the statement is true, provide an argument that proves that it is true. Use the definitions of valid and falsifiable.

(a) *If F is falsifiable and G is falsifiable, then F $\lor$ G is falsifiable.*

(b) *If F is valid and G is falsifiable, then F $\to$ G is falsifiable.*

(c) *If F $\to$ H is valid, then F $\to$ (G $\to$ H) is valid.*

(d) *If F $\to$ H is falsifiable, then F $\to$ (G $\to$ H) is falsifiable.*

5.13 Provide direct proofs of the following statements:

(a) *If (P $\to$ $\bot$) is false, then P is true.*

(b) *If (P $\lor$ Q) is true and ¬P is true, then Q is true.*

(c) *All valuations make P $\to$ (Q $\to$ P) true.*

(d) *If a number is divisible by 6, then it is also divisible by 3.*

(e) *The set (A $\cup$ B) is a subset of (A $\cup$ B) $\cup$ C.*

(f) *The set (A $\cap$ B) $\cap$ C is a subset of (A $\cap$ B).*

(g) *If A $\subseteq$ (B $\cap$ C), then A $\subseteq$ B.*

(h) *If x $\in$ A $\cup$ (B $\cap$ C), then x $\in$ (A $\cup$ B) $\cap$ (A $\cup$ C).*

5.14 We have seen that A $\land$ (B $\lor$ C) is equivalent to (A $\land$ B) $\lor$ (A $\land$ C), in other words, that $\land$ distributes over $\lor$, but how is it with $\to$ and the other connectives? Justify your answer to each of the following questions by either proving that it is so or providing a counterexample.

(a) Is A $\to$ (B $\land$ C) equivalent to (A $\to$ B) $\land$ (A $\to$ C)?

(b) Is A $\to$ (B $\lor$ C) equivalent to (A $\to$ B) $\lor$ (A $\to$ C)?

(c) Is (A $\land$ B) $\to$ C equivalent to (A $\to$ C) $\land$ (B $\to$ C)?

(d) Is (A $\lor$ B) $\to$ C equivalent to (A $\to$ C) $\lor$ (B $\to$ C)?

(e) Is A $\land$ (B $\to$ C) equivalent to (A $\land$ B) $\to$ (A $\land$ C)?

(f) Is A $\lor$ (B $\to$ C) equivalent to (A $\lor$ B) $\to$ (A $\lor$ C)?

5.15 Are there propositional formulas G and H such that (G $\to$ H) and (H $\to$ G) both are contradictions? Justify your answer.

5.16   Look at the following statements:

(a) X *is true if and only if* Y *is true.*
(b) X *is valid if and only if* Y *is valid.*

Does (a) follow from (b)? Does (b) follow from (a)? Explain.

5.17   For the following choices of connective $\square$, determine whether the propositional formula

$$(F \to G) \to ((F \square H) \to (G \square H))$$

is valid. Give reasons for your answer. If it is valid, provide an explanation, for example using a truth table. If it is not valid, provide a valuation that makes the formula false.

(a) $\square = \vee$ (b) $\square = \wedge$ (c) $\square = \to$

5.18   Discuss the differences between these statements:

(a) $(G \to H) \vee (H \to G)$ *is a valid propositional formula.*
(b) *For all propositional formulas* G *and* H, $(G \to H)$ *or* $(H \to G)$ *is a valid formula.*

Are both statements true?

5.19   Prove the following statement: "For all propositional formulas A, A is valid if and only if $\neg$A is not satisfiable."

5.20   Explain using your own intuition and your own example why the statement *if A, then B* and its contrapositive, *if not B, then not A*, are equivalent.

5.21   Show that C is a logical consequence of the set consisting of the formulas $(A \vee B)$, $(A \to C)$, and $(B \to C)$.

5.22   Show that $A \to C$ is a logical consequence of $(A \to B) \wedge (B \to C)$.

5.23   Suppose that F and G are logically equivalent. Show that G is a logical consequence of F and vice versa.

Chapter 6

Relations

■■■■■▮■■■■■■■■■■■■■■■■■■■

The aim of this chapter is to understand what relations are and to become familiar with the most important properties of relations, such as reflexivity, symmetry, transitivity, antisymmetry, and irreflexivity. You will also learn about what makes a relation an equivalence relation, a partial order, or a total order.

Abstraction over Relations

In this chapter we will study *relations* in a mathematical and abstract perspective. The terms we use for relations are found in everything from database theory and programming to linguistics and philosophy, and it is therefore worth getting to know them well. We actually meet *relations* all the time. We use the less-than-or-equal-to relation $\leqslant$ on real numbers, the subset relation $\subseteq$ on sets, and the logical consequence relation $\Rightarrow$ on propositional formulas. A list of who knows whom is also a relation. Now we will *generalize* and *abstract* over this phenomenon by *defining* what is meant by a relation. Then we will investigate what properties a relation can have.

Definition 6.1. Relation

A **binary relation** *from* a set S to a set T is a subset of $S \times T$. A **binary relation** *on* a set S is a subset of $S^2 = S \times S$. More generally, an **n-ary relation** is a subset of $A_1 \times \cdots \times A_n$, and an n-ary relation *on* a set A is a subset of A^n.

When we use the word "relation," it is usually implicitly understood that it means a *binary* relation. In order to get to know the concept of a *relation* and what the definition entails, we shall now look at many examples of relations. A relation can be represented and visualized in many different ways, and for each of the following examples we shall provide *the set of tuples* that make up the relation, we shall draw the relation as a *figure with arrows*, and we will construct a *table* of exactly those tuples that are present in the relation.

© Springer Nature Switzerland AG 2021
R. Antonsen, *Logical Methods*, https://doi.org/10.1007/978-3-030-63777-4_7

Q   **Binary relation.** The set $\{\,\langle a, 1\rangle\,, \langle a, 2\rangle\,, \langle b, 2\rangle\,\}$ is a binary relation *from* $\{a, b\}$ *to* $\{1, 2\}$ because it is a subset of $\{a, b\} \times \{1, 2\}$. Each of the following representations of this relation captures exactly the same thing, namely that a is related to 1 and 2, that b is related to 2, and that b is *not* related to 1:

$$\{\,\langle a, 1\rangle\,, \langle a, 2\rangle\,, \langle b, 2\rangle\,\}$$

	1	2
a	1	1
b	0	1

The following is a binary relation *on* the set $\{1, 2, 3\}$:

$$\{\,\langle 1, 2\rangle\,, \langle 1, 3\rangle\,, \langle 2, 3\rangle\,, \langle 3, 3\rangle\,\}$$

	1	2	3
1	0	1	1
2	0	0	1
3	0	0	1

◆

Some Special Relations

We always have the following special relations when two sets S and T are given:

The **identity relation** on S, which relates each element to itself and to nothing else: $\{\,\langle x, x\rangle \mid x \in S\,\}$. Here is *the identity relation* on the set $\{1, 2, 3\}$:

$$\{\,\langle 1, 1\rangle\,, \langle 2, 2\rangle\,, \langle 3, 3\rangle\,\}$$

	1	2	3
1	1	0	0
2	0	1	0
3	0	0	1

The **empty relation** from S to T, which does not relate any elements to each other: $\varnothing$. Here is *the empty relation* on the set $\{1, 2, 3\}$:

$$\varnothing$$

	1	2	3
1	0	0	0
2	0	0	0
3	0	0	0

The **universal relation** from S to T, which relates every element of S to every element of T: $S \times T$. Here is *the universal relation* on the set $\{1, 2, 3\}$:

$$\{\langle 1, 1\rangle\,, \langle 1, 2\rangle\,, \langle 1, 3\rangle\,, \\ \langle 2, 1\rangle\,, \langle 2, 2\rangle\,, \langle 2, 3\rangle\,, \\ \langle 3, 1\rangle\,, \langle 3, 2\rangle\,, \langle 3, 3\rangle\,\}$$

	1	2	3
1	1	1	1
2	1	1	1
3	1	1	1

Q   Several other binary relations. We already know several binary relations: the equality relation =, the subset relation ⊆, the less-than-or-equal-to relation ≤, the less-than relation <, and the logical consequence relation ⇒. We also have relations on other types of objects, for example we have the parent relation, the sibling relation, the romantic partner relation, and the friend relation, all of which are relations on the set of human beings.   ♦

Notation. If R is a relation and ⟨a, b⟩ ∈ R, we allow ourselves to write aRb instead of ⟨a, b⟩ ∈ R. For example, we write 4 < 5 instead of ⟨4, 5⟩ ∈ <. This is called **infix notation**, which you will meet again later on several occasions (*pages 122 and 152*).

The Universe of Relations

We have now defined a new mathematical object, a *relation*. The definition of a relation is very general in the sense that we can have relations between whatever sets we want, and it is very precise because it accurately states what makes up a relation and what does not. We can now start exploring sets of relations and identify interesting properties of them.

Let us look at *all* the relations on a set of two elements. There are 16 of them in total. The empty relation can be drawn as • • , without any arrows, which shows that nothing is related, and the universal relation can be drawn as ↫, with all the arrows in place, which shows that everything is related to everything. Each arrow represents exactly one tuple of the relation. Here are all possible relations, drawn in two different ways:

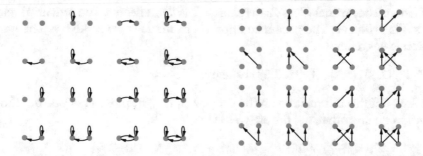

The figure to the right shows all relations from a set of two elements to *another* set of two elements. It gives rise to the figure on the left if two dots are combined into one. Make sure before reading on that you understand the figures and that you see that all the relations are actually represented. We will now look at some important properties that binary relations may have. They are *reflexivity, symmetry, transitivity, antisymmetry,* and *irreflexivity*. There are many more properties that relations may have, but these play a central role.

Reflexivity, Symmetry, and Transitivity

Definition 6.2. Reflexive relation
A binary relation R on a set S is **reflexive** if for all x in S it is the case that $\langle x, x \rangle \in R$.

Reflexive relations are those for which everything is related to itself. The only reflexive relations on a set of two elements are the following:

The relation is not reflexive, because the element on the left is not related to itself.

Q   **Reflexive relation.** Let S be the set $\{1, 2, 3\}$. Then $\{\langle 1, 1 \rangle, \langle 2, 2 \rangle, \langle 3, 3 \rangle\}$ is a reflexive relation on S because all the tuples of the form $\langle x, x \rangle$ are in the set. The same thing is true of the relation $\{\langle 1, 1 \rangle, \langle 1, 2 \rangle, \langle 2, 2 \rangle, \langle 2, 3 \rangle, \langle 3, 3 \rangle\}$. But $\{\langle 1, 1 \rangle, \langle 1, 3 \rangle\}$ is *not* a reflexive relation on S, because both $\langle 2, 2 \rangle$ and $\langle 3, 3 \rangle$ are missing. The relation $\{\langle 1, 1 \rangle, \langle 2, 2 \rangle\}$ is also not reflexive, this time because $\langle 3, 3 \rangle$ is missing.   ♦

Q   **Reflexive relations.** Let us check whether any relations that we already know are reflexive. Make sure you understand each of the following: The equality relation on a set is always reflexive. The subset relation $\subseteq$ is reflexive. The less-than-or-equal-to relation $\leqslant$ on numbers is reflexive. The logical consequence relation $\Rightarrow$ on formulas is reflexive. The ordinary less-than relation $<$ is *not* reflexive. The universal relation on a set is reflexive. The parent relation on humans is *not* reflexive. The partner relation is *not* reflexive.   ♦

?   Is $\{\langle 1, 1 \rangle, \langle 2, 2 \rangle, \langle 3, 3 \rangle, \langle 4, 4 \rangle, \langle 5, 5 \rangle\}$ reflexive?   ♦

!   It depends! If this is a relation on the set $\{1, 2, 3, 4, 5\}$, the answer is yes, but if this is a relation on any other set, the answer is no.   ♦

Definition 6.3. Symmetric relation
A binary relation R on a set S is **symmetric** if for all x, y it is the case that if $\langle x, y \rangle \in R$, then $\langle y, x \rangle \in R$.

Symmetric relations are those for which the order of elements does not matter. The only symmetric relations on a set with two elements are the following:

The relation  is not symmetric, because the element on the left is related to the one on the right, but not the other way around.

Q **Symmetric   relation.** Let   S   be   the   set $\{a, b, c\}$.   Then   the   relation $\{\, \langle a, b \rangle, \langle b, a \rangle, \langle c, b \rangle, \langle b, c \rangle \,\}$ is a symmetric relation on S, but $\{\, \langle a, b \rangle, \langle b, a \rangle, \langle a, c \rangle \,\}$ is *not* symmetric, because $\langle a, c \rangle$ is present without $\langle c, a \rangle$ also being present.   ◆

Q **Symmetric relations.** Let us check whether any relations that we already know are symmetric. Make sure you understand each of the following: The equality relation on a set is always symmetric. The subset relation $\subseteq$ is *not* symmetric. The less-than-or-equal-to relation $\leqslant$ on numbers is *not* symmetric. The logical consequence relation $\Rightarrow$ on formulas is *not* symmetric. The logical equivalence relation $\Leftrightarrow$ on formulas is symmetric. The universal relation on a set is symmetric. The parent relation on humans is *not* symmetric. The partner relation is symmetric.◆

Definition 6.4. Transitive relation

A binary relation R on a set S is **transitive** if for all x, y, z it is the case that if $\langle x, y \rangle \in R$ and $\langle y, z \rangle \in R$, then $\langle x, z \rangle \in R$.

Transitive relations are those for which everything you can do in two steps, you can also do in one step.

The relation  on a set of three elements is not transitive, because the element on the far left is not related to the element on the far right. On the other hand, the relation  is transitive, because all the required arrows are present. Similarly, the relation  is not transitive, because the left element is not related to itself.

Q **Transitive relations.** Let S be the set $\{1, 2, 3\}$. Then the relation $\{\, \langle 1, 2 \rangle, \langle 2, 3 \rangle, \langle 1, 3 \rangle \,\}$ is a transitive relation on S: we have that both $\langle 1, 2 \rangle$ and $\langle 2, 3 \rangle$ are present, and the requirement of transitivity says that then $\langle 1, 3 \rangle$ must also be present, which is the case. But $\{\, \langle 1, 2 \rangle, \langle 2, 3 \rangle \,\}$ is *not* a transitive relation, because $\langle 1, 3 \rangle$ is missing. The relation $\{\, \langle 1, 2 \rangle, \langle 2, 1 \rangle \,\}$ is also not transitive, and that is because $\langle 1, 2 \rangle$ and $\langle 2, 1 \rangle$ are present, but both $\langle 1, 1 \rangle$ and $\langle 2, 2 \rangle$ are missing.   ◆

Q **Transitive relations.** Let us check whether any relations that we already know are transitive. Make sure you understand each of the following: The equality relation on a set is always transitive. The subset relation $\subseteq$ is transitive. The less-than-or-equal-to relation $\leqslant$ on numbers is transitive. The logical consequence relation $\Rightarrow$ and equivalence relation $\Leftrightarrow$ on formulas are transitive. The parent relation on humans is *not* transitive, but the ancestor relation is transitive. The friend relation is *not* transitive. The universal relation on a set is transitive. The partner relation is *not* transitive.   ◆

Definition 6.5. Equivalence relation

A binary relation on a set S is an **equivalence relation** if it is reflexive, symmetric, and transitive.

Equivalence relations are those that tell us whether elements are in some sense "equal." The only equivalence relations on a set with two elements are the following:

The relation 💋 is not an equivalence relation, because it is not symmetric.

Q   **Equivalence relation.**   Let S be the set $\{1, 2, 3\}$. Then the relation

$$\{\, \langle 1, 1 \rangle \,,\, \langle 2, 1 \rangle \,,\, \langle 1, 2 \rangle \,,\, \langle 2, 2 \rangle \,,\, \langle 3, 3 \rangle \,\}$$

is an equivalence relation on S, because it is reflexive, symmetric, and transitive. But the relation $\{\, \langle 1, 1 \rangle \,,\, \langle 2, 1 \rangle \,,\, \langle 1, 2 \rangle \,\}$ is not an equivalence relation on S, because it is neither reflexive nor transitive.   ◆

Antisymmetry and Irreflexivity

Equivalence relations behave nicely, but the picture is not complete until we look at *antisymmetry* and *irreflexivity*. These properties are *not* exactly the opposites of symmetry and reflexivity, but almost. Once we have these properties in place, we can explore other types of relations.

Definition 6.6. Antisymmetric relation

A binary relation R on a set S is **antisymmetric** if for all x, y it is the case that if $\langle x, y \rangle \in R$ and $\langle y, x \rangle \in R$, then $x = y$.

Antisymmetric relations are those for which there do not exist two distinct elements each of which is related to the other. The only relations on a set of two elements that are *not* antisymmetric are those for which each element is related to the other:

Digression

Modal logic is a logic that can be used to represent and analyze *relational structures*. Modal logic extends propositional logic with the **modal operators** □ and ◇. A common interpretation of □ and ◇ is that ◇P represents "it is possible that P" and that □P represents "it is necessary that P." The operators can be expressed by means of each other and negation. For example, ¬□¬P expresses the same thing as ◇P: "it is not necessary that you do not get the prize" and "it is possible that you get the prize" mean the same thing. It is common to interpret formulas in modal logic using relations. Imagine a relation as a collection of points with arrows between then. At each point there is something that is true, and what is true may vary from point to point. Imagine that you are at one of these points and that you are looking at which points you can get to by following one of the arrows. Call these points the *accessible points* from where you are. Formulas in modal logic may now be interpreted as follows: The formula ◇P is true at a point if there is an accessible point where P is true. The formula □P is true at a point if P is true at all accessible points. There is a beautiful correspondence between formulas in modal logic and properties of relations. For example, the formula P → ◇P is true for all reflexive relations, ◇◇P → ◇P is true for all transitive relations, and P → □◇P is true for all symmetric relations. Modal logic may be used to represent propositions about time in temporal logic, about knowledge in epistemic logic, about duty in deontic logic, and several other phenomena. In theoretical computer science modal logic is an essential tool for proving properties of programs.

Q Antisymmetry. Let S be the set $\{1, 2, 3\}$. The relation $\{\langle 1, 2 \rangle, \langle 1, 3 \rangle\}$ is antisymmetric, but $\{\langle 1, 2 \rangle, \langle 2, 1 \rangle\}$ is not antisymmetric, because $\langle 1, 2 \rangle$ and $\langle 2, 1 \rangle$ are present, but $1 \neq 2$. ◆

Definition 6.7. Irreflexive relation

A binary relation R on a set S is **irreflexive** if there is no $x \in S$ such that $\langle x, x \rangle \in R$.

Irreflexive relations are those for which nothing is related to itself. The only irreflexive relations on a set of two elements are the following:

The relation ⟿ is not irreflexive, because the element to the left is related to itself.

Q **Irreflexivity.** Let S be the set $\{1, 2, 3\}$. The relation $\{\langle 1, 2\rangle, \langle 2, 1\rangle\}$ is an irreflexive relation on S, but $\{\langle 3, 1\rangle, \langle 3, 2\rangle, \langle 3, 3\rangle\}$ is not, because $\langle 3, 3\rangle$ is present.   ◆

Note. Both reflexivity and irreflexivity are universal statements. In a reflexive relation *all* tuples of the form $\langle x, x\rangle$ are present. In an irreflexive relation *no* tuples of the form $\langle x, x\rangle$ are present. Thus being irreflexive does not mean the same as *not* being reflexive. In a relation that is not reflexive, at least one tuple $\langle x, x\rangle$ is missing, while in an irreflexive relation, *all* tuples $\langle x, x\rangle$ are missing.

Orders, Partial and Total

We have an intuitive understanding of what an *order* is and what it means to *order* a set, but with the terminology we have now defined, we can make this precise in the following way.

Definition 6.8. Partial order

A binary relation on a set S is a **partial order** if it is reflexive, transitive, and anti-symmetric.

Partial orders are relations with "direction." The only partial orders on a set with two elements are the following:

Notice that the only difference between this definition and the definition of an equivalence relation is that "symmetric" has been replaced with "antisymmetric." The reason for the word "partial" is that there may be two elements that are not related to each other. The simplest example of this is the relation , in which the two elements are related to themselves, but not to each other.

Q **Partial orders.** The less-than-or-equal-to relation $\leqslant$ on real numbers is a partial order, but the less-than relation $<$ is *not*, because it is not reflexive. The logical consequence relation $\Rightarrow$ is also not a partial order, because it is not antisymmetric: for example, we have that $P \Rightarrow \neg\neg P$ and $\neg\neg P \Rightarrow P$, but P and $\neg\neg P$ are not identical.◆

Q **The subset relation.** If we have a set of sets, the subset relation $\subseteq$ on this set is always a partial order: It is *reflexive* because $X \subseteq X$ for all sets X, it is *transitive* because $X \subseteq Y$ and $Y \subseteq Z$ always implies $X \subseteq Z$, and it is *antisymmetric* because $X \subseteq Y$ and $Y \subseteq X$ always implies $X = Y$. We may, for example, look at all the subsets of $\{1, 2, 3\}$. There are exactly eight of them, and $\subseteq$ is a partial order on the set of them. The smallest set is the empty set, and the largest set is $\{1, 2, 3\}$. We may illustrate the situation like this:

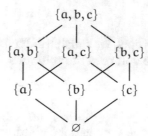

Here we have drawn neither the arrows that correspond to the relation being *reflexive* nor those that follow from the fact that the relation is *transitive*, but these may be read out of the diagram: if we can go from X *up* to Y by following the lines, we have that $X \subseteq Y$.   ♦

The diagram in this example is called a **Hasse diagram**, named after the German mathematician *Helmut Hasse* (1898–1979). What characterizes Hasse diagrams is that they contain just enough lines to unambiguously represent a partial order. Even though there is *not* a line directly from $\{a\}$ to $\{a, b, c\}$, the diagram says that $\{a\} \subseteq \{a, b, c\}$ because we can go from $\{a\}$ *up* to $\{a, b, c\}$ by following the lines. Hasse diagrams generally enable us to see the underlying structure of a partial order clearly.

Definition 6.9. Total order

A partial order R on a set S is called a **total order** or a **linear order** if for all x and y in S it is the case that xRy or yRx.

Total orders are relations that arrange all the elements of a set one after the other. The only total orders on a set of two elements are

When an order is total, all the elements are related to each other, and that means we can place them next to each other on a line.

Q   **Total order.** The relation $\leqslant$ on real numbers is a total order. The subset relation does not need to be a total order, because it is possible to have two sets A and B with neither $A \subseteq B$ nor $B \subseteq A$.   ♦

Examples

In order to master these concepts, we finish this chapter with some exercises. Try to solve them yourself before you read the solutions.

❓ Which properties does the "is a child of" relation have on the set of all human beings? ◆

❗ *Reflexive?* No, there is someone who is not a child of him- or herself. *Symmetric?* No, "X is a child of Y" does not imply that "Y is a child of X." *Transitive?* No, "X is a child of Y" and "Y is a child of Z" does not give that "X is a child of Z." *Antisymmetric?* Yes, because "X is a child of Y" and "Y is a child of X" are never true simultaneously. *Irreflexive?* Yes, no one is their own child. *Equivalence relation?* No, it is neither reflexive nor symmetric nor transitive. *Partial order?* No, it is neither reflexive nor transitive. *Total order?* No, it is not even a partial order. ◆

❓ Which properties does the greater-than-or-equal-to relation $\geqslant$ have on the set of real numbers? ◆

❗ *Reflexive?* Yes, all real numbers equal themselves. *Symmetric?* No, we have $3 \geqslant 2$, but it is not the case that $2 \geqslant 3$. *Transitive?* Yes, if $x \geqslant y$ and $y \geqslant z$, then $x \geqslant z$. *Antisymmetric?* Yes, if $x \geqslant y$ and $y \geqslant x$, then $x = y$. *Irreflexive?* No, we have that $1 \geqslant 1$. *Equivalence relation?* No, it is not symmetric. *Partial order?* Yes, it is reflexive, transitive, and antisymmetric. *Total order?* Yes, if we take two numbers, one of them will always be greater than or equal to the other. ◆

❓ Which properties does the relation ⌒→ have? ◆

❗ *Reflexive?* No, because the arrows ↻ ↻ ● are missing. *Symmetric?* No, because the arrows ●⌒● are missing. *Transitive?* No, because we have ⌒⌢, but not ●⌒ ●. *Antisymmetric?* Yes. *Irreflexive?* No, because we have ● ● ↻. *Equivalence relation?* No, it is neither transitive nor reflexive nor symmetric. *Partial order?* No, it is neither reflexive nor transitive. *Total order?* No, it is not even a partial order. ◆

❓ Which properties does the relation ↻⌣↻ have? ◆

❗ *Reflexive?* Yes, because all the elements are related to themselves. *Symmetric?* No, because ●⌒● is missing. *Transitive?* Yes. *Antisymmetric?* Yes. *Irreflexive?* No, because we have ↻ ↻ ↻. *Equivalence relation?* No, because it is not symmetric. *Partial order?* Yes. *Total order?* Yes. ◆

Exercises

6.1 Here are seven binary relations on the set $\{1, 2, 3\}$. For each of them, determine which of the following properties it has: *reflexive, symmetric, transitive, antisymmetric,* and *irreflexive*. Hint: It may be a good idea to draw the relations.

(a) $R_1 = \{\, \langle 1, 1 \rangle, \langle 2, 2 \rangle, \langle 3, 3 \rangle \,\}$

(b) $R_2 = \{\, \langle 1, 1 \rangle, \langle 2, 2 \rangle, \langle 3, 3 \rangle, \langle 1, 2 \rangle \,\}$

(c) $R_3 = \{\, \langle 1, 1 \rangle, \langle 2, 2 \rangle, \langle 3, 3 \rangle, \langle 1, 2 \rangle, \langle 2, 3 \rangle \,\}$

(d) $R_4 = \varnothing$

(e) $R_5 = \{\, \langle 1, 2 \rangle \,\}$

(f) $R_6 = \{\, \langle 1, 1 \rangle, \langle 1, 2 \rangle, \langle 2, 1 \rangle \,\}$

(g) $R_7 = \{\, \langle 3, 3 \rangle, \langle 3, 2 \rangle, \langle 2, 1 \rangle \,\}$

6.2 In this exercise you will show that the three properties reflexive, symmetric, and transitive are *independent* of each other. For each item, find a relation that has all the following properties:

(a)   reflexive,   symmetric, and   transitive

(b)   reflexive,   symmetric, and *not* transitive

(c)   reflexive, *not* symmetric, and   transitive

(d)   reflexive, *not* symmetric, and *not* transitive

(e)   *not* reflexive,   symmetric, and   transitive

(f)   *not* reflexive,   symmetric, and *not* transitive

(g)   *not* reflexive, *not* symmetric, and   transitive

(h)   *not* reflexive, *not* symmetric, and *not* transitive

6.3 Here are some binary relations on the set $\{a, b, c\}$.

- $S_1 = \{\, \langle a, a \rangle, \langle a, b \rangle, \langle b, b \rangle, \langle b, c \rangle \,\}$
- $S_2 = \{\, \langle a, a \rangle, \langle a, b \rangle, \langle b, b \rangle \,\}$
- $S_3 = \{\, \langle a, a \rangle, \langle a, b \rangle, \langle b, c \rangle \,\}$
- $S_4 = \{\, \langle a, a \rangle, \langle a, b \rangle, \langle b, a \rangle, \langle b, b \rangle, \langle c, c \rangle \,\}$
- $S_5 = \{\, \langle a, a \rangle, \langle b, c \rangle, \langle c, b \rangle \,\}$

(a) Which element is S_1 missing in order for it to be a *reflexive* relation?

(b) Which element is S_2 missing in order for it to be a *symmetric* relation?

(c) Which element is S_3 missing in order for it to be a *transitive* relation?

(d) Which element may be removed from S_4 in order to make it *antisymmetric*?

(e) Which element may be removed from S_5 in order to make it *irreflexive*?

(f) Which of the relations are *reflexive*? *symmetric*? *transitive*?

6.4 Let $A = \{a, b, c, d, e\}$, and let R be a relation on A defined as follows:

$$R = \{\, \langle a, c \rangle, \langle a, d \rangle, \langle b, a \rangle, \langle b, d \rangle, \langle c, e \rangle, \langle e, b \rangle, \langle e, e \rangle \,\}.$$

Draw R and find out what properties it has.

6.5 If d and n are natural numbers, we say that d **divides** n if there is a natural number m such that dm = n. For example, it is the case that 3 divides 6, because there is a natural number m, in fact 2, such that 3m = 6. Let A = {2, 3, 4} and B = {6, 8, 10}. Let R be the relation from A to B defined by letting xRy if x divides y.

  (a) Is ⟨4, 6⟩ ∈ R?   (b) Is ⟨4, 8⟩ ∈ R?   (c) Is ⟨3, 8⟩ ∈ R?   (d) Is ⟨2, 10⟩ ∈ R?

Write R as a set of ordered pairs.

6.6 (a) Can a relation be both symmetric and antisymmetric? If yes, provide an example; if no, explain why not.

  (b) Can a relation be both reflexive and irreflexive? If yes, provide an example; if no, explain why not.

6.7 (a) Suppose that R and S are equivalence relations, and let T = R ∩ S. Show that T is also an equivalence relation.

  (b) Show that it is *not* always the case that if R and S are equivalence relations, then R ∪ S is also an equivalence relation.

6.8 Let ~ be a relation on $\mathbb{N} = \{0, 1, 2, 3, \dots\}$ defined by letting $x \sim y$ if $x + y$ is an even number. Which properties does this relation have?

6.9 Let ~ be a relation on $\mathbb{N} = \{0, 1, 2, 3, \dots\}$ defined by letting $x \sim y$ if $x \cdot y$ is an even number. Which properties does this relation have?

6.10 A relation R is **asymmetric** if for all x and y such that xRy is true, yRx is false. Is it true that a relation is asymmetric if and only if it is antisymmetric? Explain.

6.11 Two sets A and B are called **disjoint** when they do not have any elements in common, that is, when $A \cap B = \emptyset$.

  (a) Show that every set is disjoint from the empty set.

  (b) Let A, B, and C be three distinct sets. Suppose that A and B are disjoint, and that B and C are disjoint. Show that then A and C are also disjoint, or provide a counterexample that shows that this is not the case.

  (c) Let ~ be a relation such that A ~ B is true if A and B are disjoint. Is this relation an equivalence relation?

6.12 A relation R on a set U is called a **preorder** on U if it is reflexive and transitive.

  (a) Show that ⇒ is a preorder.

  (b) Show that ⇒ is *not* a partial order.

  (c) Suppose that ⩽ is a preorder on U. Let ~ be a relation on U such that $x \sim y$ if and only if $x \leqslant y$ and $y \leqslant x$. Prove that ~ is an equivalence relation.

Chapter 7

Functions

■■□□■■□■□■■■□■■■■■■■■■□■■■

In this chapter you will learn about what functions are, as well as the most important concepts related to functions, such as whether they are injective, surjective, or bijective. You will also learn about functions with multiple arguments, how functions can be composed, what operations are, and how functions can be objects.

What Is a Function?

– What do you say if I say 4?	*– Then I say 13.*
– But what if I say 5?	*– Then I have to say 16.*
– And if I say 1?	*– Then I have to say 4.*
– What if I say 4 again?	*– Then I have to say 13 again.*
– Hmmm, what about 0?	*– Then I have to say 1.*
– Exactly!	*– Can you guess what I do?*

This dialogue illustrates the most important thing about a function: no matter what number is said – this is called the *argument* of the function – there is exactly one answer – this is called the *value* of the function. We will now define what a function is in an abstract and precise way, as well as look at some important properties of functions. Most people know functions on numbers, for example, the function that takes a number as an argument and returns triple the number as a value. We will see that functions do not have to be on numbers, but can be on *anything*. If we have two sets, and *each* element of one set is associated with *exactly* one element of the other set, we have a *function* between the sets:

Definition 7.1. Function

A **function** from A to B is a binary relation f from A to B such that for *every* $x \in A$, there is *exactly* one element $y \in B$ such that $\langle x, y \rangle \in f$. We write $f(x) = y$ when $\langle x, y \rangle \in f$. In this case, we call x the **argument** and $f(x)$ the **value** of the function. We write $f : A \to B$ for the function f when it is a function from A to B.

© Springer Nature Switzerland AG 2021
R. Antonsen, *Logical Methods*, https://doi.org/10.1007/978-3-030-63777-4_8

We can think of a function $f : A \to B$ as a box, a machine, or a program with an input and an output. If we put x in, we get $f(x)$ out. It behaves in such a way that no matter what we put in from A, we get out an element from B, and if we put in the same element from A again, we get the same element from B again. Sometimes we are interested in the inner workings of the box; at other times, we are interested only in what it does. The functions $f(x) = x + x$ and $g(x) = 2x$, for example, do exactly the same thing, but they are defined differently. We say that they are *extensionally* equal, but *intensionally* different.

A function can be viewed as a box or machine with an input and an output.

Notation. It is common to write, for example, $f(x) = 2x$ for the function that contains all the tuples of the form $\langle x, 2x \rangle$. Another way of writing the same function is $x \mapsto 2x$. The expression $f(x) = y$ can be read in several ways. Some of the more common of these are "f of x equals y," "f maps x to y," and "f applied to x gives/returns y." The expression $f(x)$ is used in several ways. It usually means the *result* of applying the function f to the argument x, but it is also used to symbolize the actual function.

Q   **A simple function.** The relation $\{ \langle 1, 1 \rangle, \langle 2, 2 \rangle, \langle 3, 3 \rangle \}$ is a function from $\{1, 2, 3\}$ to $\{1, 2, 3\}$. If we call this function f, we get that $f(1) = 1$, $f(2) = 2$, and $f(3) = 3$.   ◆

Q   **Not a function.** The relation $\{ \langle 1, b \rangle, \langle 2, a \rangle \}$ from $\{1, 2, 3\}$ to $\{a, b\}$ is *not* a function, because it lacks a tuple of the form $\langle 3, x \rangle$. The relation $\{ \langle 1, 1 \rangle, \langle 1, 2 \rangle \}$ on $\{1, 2\}$ is also not a function, because 1 occurs as the first element in two tuples. That is not allowed.   ◆

Definition 7.2. Domain and codomain

Let f be a function from A to B. The set A is called the **domain** of f, and the set B is called the **codomain** of f.

There are two essential properties that a function must have in addition to being a binary relation, and both properties can be formulated using the terms *domain* and *codomain*: First, a function cannot send an element of the domain to two *different* elements of the codomain. Second, a function must map *each element* of the domain to an element of the codomain. If a relation has these properties, it is a function.

Examples of binary relations that are not functions.

Q **Assignments and valuations.** An assignment of truth values to propositional variables is a function from the set of propositional variables to $\{0, 1\}$. In Chapter 3 we defined a *valuation* as a function from the set of all propositional formulas to $\{0, 1\}$, with the requirement that the function be in agreement with the tables in Definition 3.2 (*page 30*). There are many functions from propositional formulas to $\{0, 1\}$ that are *not* valuations, for example the function that maps all formulas to **1**.◆

Q **The successor function.** The function $s : \mathbb{N} \to \mathbb{N}$ such that $s(x) = x + 1$ is called the **successor function** on the natural numbers. This function takes a natural number as argument and returns the successor of this number, that is, the next natural number. ◆

The simplest function of all is the function that does not do anything, but just maps x to x:

Definition 7.3. The identity function

If A is a set, the **identity function** on A is the function id_A such that $id_A(x) = x$ for all $x \in A$.

Q **A function for even and odd numbers.** The function $\text{Even} : \mathbb{N} \to \{0, 1\}$ defined by

$$\text{Even}(x) = \begin{cases} 1 \text{ if } x \text{ is an even number} \\ 0 \text{ if } x \text{ is an odd number} \end{cases}$$

has $\mathbb{N}$ as its domain and $\{0, 1\}$ as its codomain, and can be used to identify the even and odd numbers. ◆

Injective, Surjective, and Bijective Functions

Now we will look at some basic properties of functions that enable us to categorize functions in a natural and useful way. The first property is called *injectivity*. It means that *different* elements of the domain are always mapped to *different* elements of the codomain. The other property is called *surjectivity*. A function is surjective if every element of the codomain is "hit" by the function. Now we will define these two properties precisely.

Definition 7.4. Injective

A function $f : A \rightarrow B$ is **injective** if for all elements x and y in A, if $x \neq y$, then $f(x) \neq f(y)$. In this case we say that f is an **injection** or **one-to-one**.

Injectivity therefore means that two *different* elements are mapped to two *different* elements.

Example of two injective functions to the left and two noninjective functions to the right.

Q   **injective.** The function $f(x) = 2x$ from $\mathbb{R}$ to $\mathbb{R}$ is injective. The function $f(x) = x^2$ from $\mathbb{R}$ to $\mathbb{R}$ is *not* injective, because $1 \neq -1$, but $f(1) = f(-1)$.   ◆

In order to prove that a function is injective, it is often easier to prove the contrapositive statement. The contrapositive statement is that a function $f : A \rightarrow B$ is injective if for all elements x and y in A, it is the case that if $f(x) = f(y)$, then $x = y$.

?   Prove that the function $f(x) = 2x$ is injective.   ◆

!   Suppose that $f(x) = f(y)$. Then $2x = 2y$. By dividing by 2 on both sides we get that $x = y$.   ◆

Definition 7.5. Range/image

Let f be a function from A to B, and let X be a subset of A. The set $\{f(x) \mid x \in X\}$ is called the **image** of X under f, and is written $f[X]$. The image of the entire set A under f, $f[A]$, is called the **range** of f.

The range of a function is what is "being hit" by the function, and a natural question is whether the range is all of the codomain. If that is the case, we say that the function is *surjective*, or *onto*. The next definition makes this precise.

Definition 7.6. Surjective

A function $f : A \rightarrow B$ is **surjective** if for all $y \in B$, there is an $x \in A$ such that $f(x) = y$. In that case, we say that f is a **surjection** or f is **surjective** or f is **onto**.

A surjective function is a function that hits everything in the codomain; that is, the codomain is identical to the range.

Example of two surjective functions to the left and two nonsurjective functions to the right.

Q **Surjective.** The function $f(x) = 2x$ from $\mathbb{R}$ to $\mathbb{R}$ is surjective. The function $f(x) = x^2$ from $\mathbb{R}$ to $\mathbb{R}$ is *not* surjective, because -1 is in the codomain, but there is no x in the domain such that $f(x) = -1$.   ◆

To prove that a function f is a surjection, we must show that for *each* element y in the codomain, there is an element x of the domain such that $f(x) = y$.

❷ Prove that the function $f(x) = 2x$ is surjective.   ◆

❶ Choose an arbitrary element y in $\mathbb{R}$. The goal is to show that there is an x in $\mathbb{R}$ such that $f(x) = y$. By letting x be $y/2$, we can do that. We have that $f(y/2) = 2(y/2) = y$. ◆

Now we can join the properties injective and surjective; we then get a property that is important.

> **Definition 7.7. Bijective**
>
> A function is **bijective** if it is injective and surjective. We also say that function is a **bijection** or a **one-to-one correspondence**.

Example of two bijective functions to the left and two non-bijective functions to the right.

We can use painting a wall as a metaphor: A surjective way to paint a wall is to cover the entire wall with paint. An injective way to paint a wall is not painting in the same place twice.

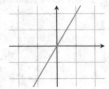

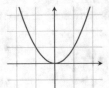

The function 2x is injective and surjective.   *The function x² is neither injective nor surjective.*

The next example shows that it is important to know what the domain and codomain of a function are.

? Let f be the function that multiplies a number by itself, that is, the function $f(x) = x^2$. *Is this function a bijection?* ◆

! It depends! It depends on the domain and the codomain. If f is a function from real numbers to real numbers, $f : \mathbb{R} \to \mathbb{R}$, it is neither injective nor surjective. Make sure that you understand why that is the case. If the domain and the codomain are each the set of real numbers greater than or equal to 0, it is both injective and surjective, and thus bijective. If we choose a positive real number y, there is *one and only one* positive real number x such that $f(x) = x^2 = y$, namely $\sqrt{y}$, the positive square root of y. ◆

Functions with Multiple Arguments

Until now we have looked only at functions that take *one* argument, but it is natural to think about functions that take more than one argument. For example, most of the ordinary calculating operations are functions that take two arguments. One way to define multiple arguments is by letting the domain be a *Cartesian product*. The multiplication function on the natural numbers can then be defined as a function from $\mathbb{N} \times \mathbb{N}$ to $\mathbb{N}$.

Notation. When the domain of a function f is a Cartesian product of n sets, we write $f(x_1, \ldots, x_n)$ instead of $f(\langle x_1, \ldots, x_n \rangle)$.

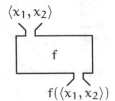

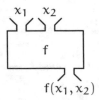

A function with a Cartesian product as its domain may be viewed as a function that takes several arguments.

The Universe of Functions

It is both interesting and educational to look at the set of *all* functions that can be defined on a set or between two sets, not only to get a better understanding of the concept of a set, but also to see how fast the complexity increases. The following figures are all based on the sets $\{1, 2\}$ and $\{1, 2, 3\}$. See whether you can identify the injective, surjective, and bijective functions.

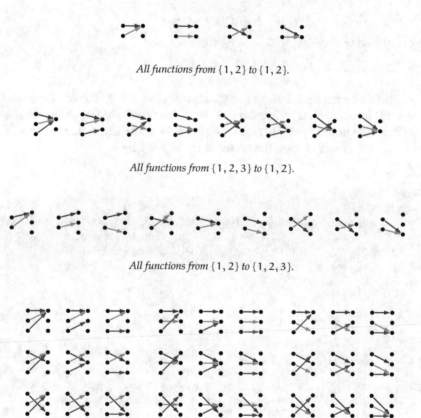

All functions from $\{1, 2\}$ to $\{1, 2\}$.

All functions from $\{1, 2, 3\}$ to $\{1, 2\}$.

All functions from $\{1, 2\}$ to $\{1, 2, 3\}$.

All functions from $\{1, 2, 3\}$ to $\{1, 2, 3\}$.

Composition of Functions

If we have two functions, we can combine them by taking the value of the first function and using it as an argument to the second function, thereby creating a new function. We can express it with figures as follows:

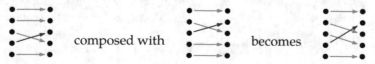

composed with   becomes

Definition 7.8. Composition of functions

If $f : A \to B$ and $g : B \to C$ are functions, the function

$$g \circ f : A \to C$$

is defined as the function we get by first applying f and then applying g to the result, that is,

$$(g \circ f)(a) = g(f(a)).$$

This new function is called the **composition** of f and g.

Q   Composition of functions. If $f(x) = 2x$ and $g(x) = x + 3$, the composition $g \circ f$ becomes the function that returns $2x + 3$. First x is multiplied by 2; then 3 is added. The composition in the opposite order, $f \circ g$, equals the function that returns $2(x+3)$. First 3 is added to x, and then the result is multiplied by 2.   ◆

Digression

A **fractal** is a mathematical object that is similar to itself when scaled. The word "fractal" was used for the first time in 1975 by the French-American mathematician *Benoît Mandelbrot* (1924–2010) and comes from the Latin word "fractus," which means fragmented or broken. Let us experiment a little. Let f be a function that works in the following way on a line:

$$f(\underline{\quad}) = \underline{\wedge}$$

If we apply this function to a composite line, like _⋀_, we get

$$f(\underline{\wedge}) = \sim\!\curlyvee\!\sim$$

Now we can repeat this process. Each line _____ that arises must be replaced with a composite line _⋀_. If we start with a triangle and repeat the process indefinitely, we get the famous **Koch snowflake,** named after the Swedish mathematician *Helge von Koch* (1870–1924):

Fractals behave strangely. This figure has a limited area, but an unlimited circumference, because if the circumference is x in a figure, it is $\frac{4}{3}x$ in the next step. The Koch snowflake will thus have a finite area, but an infinitely long circumference. We may study and make this precise via the concept of a **fractal dimension**. In this case we get the dimension $\frac{\log 4}{\log 3}$. For those who are interested, the curve is also continuous, but nowhere differentiable.

Operations

There is a particular set of functions that it is useful to be able to identify, and those are the so-called *operations*. What makes an operation special is that it does not take us "outside" a set. When we apply an operation to an element (or tuple of elements) of a domain, the result is still an element of the same set.

Definition 7.9. Operation

A **unary operation** on a set A is a function from A to A. A **binary operation** on a set A is a function from $A \times A$ to A. More generally, an **n-ary operation** on a set A is a function from A^n to A.

Q **Operations.** Whether a function is an operation depends on the domain. Addition and multiplication are operations on most sets of numbers, but such is not the case for subtraction, nor for division. For example, subtraction is not an operation on the set of natural numbers. That is because $x - y$ can be a negative number, which is outside of the domain. Similarly, division is not an operation on the set of natural numbers, because x/y can be a number that is not a natural number. On the set of real numbers, subtraction is an operation, and on the set of real numbers with 0 removed, division is an operation. In set theory, both intersection and union are operations. The successor function, which takes a natural number and adds one, is an example of a unary operation on the set of natural numbers. ♦

Functions as Objects

Functions are mathematical objects in the same way that numbers, sets, formulas, and characters are. We can compare functions, create sets of functions, define operations on them, and much more. In this chapter, we have seen how we may define basic properties of functions, such as injectivity, surjectivity, and bijectivity.

There is no reason for functions not to be able to take other functions as arguments. When they do, they are called **higher-order functions** or **functionals**. We have already seen an example of this, and that is the composition function ∘, which combines two functions into one. Another example, from mathematical analysis, is the function that takes a differentiable function f as its argument and returns the derivative of f as its value.

We can also use functions as a tool for representing other types of structures. The function defined by $f(1) = a$, $f(2) = b$, and $f(3) = c$, is one way of representing the tuple $\langle a, b, c \rangle$. The same method can also be used to represent "infinite tuples" if you wish. An "infinite tuple" with elements from a set M may be defined as a function from the natural numbers to M. In order to find the nth element of the tuple, we can apply the function to n.

> ### Digression
>
> Lambda calculus, or λ-calculus, is a formal system and a model for computability, which originated in the work of the American mathematician *Alonzo Church* (1903–1995). In λ-calculus, functions are represented syntactically by means of so-called lambda terms. For example, the expression $\lambda x.x$ represents the identity function, and $\lambda x.x+2$ represents a function that takes one argument x and returns x+2. To go from the expression x+2 to $\lambda x.x+2$ is called *lambda abstraction*. If we want to *apply* this function to the argument 4, we write $(\lambda x.x + 2)4$, that is, we put the expressions next to each other. This represents the *result* of applying the function to the argument. This expression can thereafter be *rewritten* such that we get $4 + 2$. This is another good example that something may be represented syntactically in a good way. In normal mathematical practice, the expression $f(x)$ is ambiguous; it can represent both the *function* and the *result* of applying the function to an argument. In λ-calculus this becomes more precise: the expression $\lambda x.x + 2$ represents the function, and the expression $(\lambda x.x + 2)4$ represents the result of applying the function to the argument 4. It is also easy to represent *higher-order* functions in λ-calculus: the expression $\lambda f.\lambda x.f(f(x))$ represents a function that takes one argument f and returns a function as its value. This function takes one argument and applies f to it twice. We find mechanisms and terminology from λ-calculus in many, especially functional, programming languages.

Partial Functions

In our definition of *function* there is a requirement that *each* element of the domain be mapped to an element of the codomain. But in several contexts, and especially when it comes to programs, algorithms, and computations, it is useful to relax this requirement a bit. We therefore say that a **partial function** from A to B is a function from a *subset* of A to B. The reason this is a useful term when it comes to programs and computations is that we do not always know whether a given program or algorithm will provide a value for certain elements of a domain. In other words, we do not always know that our programs provide *functions*, strictly speaking, but we still want to be able to speak accurately and prove statements about them. The procedure in Collatz's conjecture, which we met in Chapter 5 (*page 52*), is such an example, because we can define a *partial function* that gives 1 for all arguments for which the procedure terminates with the number 1, but we do not know whether this is actually a true *function*.

Exercises

7.1 For each of the properties below, define a function that has that property. Choose domains and codomains among the sets $A = \{1, 2, 3, 4\}$ and $B = \{a, b, c, d\}$ and $C = \{x, y, z\}$.

   (a) Injective, but not surjective.
   (b) Surjective, but not injective.
   (c) Injective and surjective, but not the identity function.
   (d) Neither injective nor surjective.

7.2 If the range of a function equals the codomain of the function, is it then the case that the function is surjective? Is it the case that it is injective?

7.3 (a) Give an example of a function from the set of propositional formulas to $\{0, 1\}$ that is *not* a valuation. Explain why the function is not a valuation.

   (b) Is it possible to create a valuation that is injective? Find an example or show that this is not the case.

   (c) Are all valuations surjective? Show that this is the case or find a counterexample.

   (d) Let f be an assignment of truth values to propositional variables that is *not* surjective. Describe the function f with words.

   (e) Let v be a *valuation* such that $v(X) = f(X)$ for all propositional variables X. What is $v(P \rightarrow Q)$?

7.4 Let $A = \{a, b, c, d, e, f, g, h, i\}$ and $B = \{1, 2, 3\}$.

   (a) Define a function f from A to B.
   (b) Define a *surjective* function g from A to B.
   (c) Is there an *injective* function from A to B? Justify your answer.

Let g be the surjective function you defined earlier in the exercise. We can define a binary relation R on A as follows:

$$R = \{\, \langle x, y \rangle \in A \times A \mid g(x) = g(y) \,\}.$$

That is, if x and y are elements of A, they are related if they are mapped to the same element of B by the function g. Write out the relation R as a set of pairs and answer the following questions.

   (d) Is R reflexive?   (g) Is R transitive?
   (e) Is R symmetric?   (h) Is R an equivalence relation?
   (f) Is R antisymmetric?   (i) Is R a partial order?

Will the answers to these questions be different if we choose another g?

7.5   Let the function $f : \mathbb{N} \to \mathbb{N}$ be defined by $f(x) = 2x + 1$. Let O be the set of odd numbers, let E be the set of even numbers, and describe the following sets.

(a) The range of f   (c) f[O]

(b) f[E]   (d) f[∅]

7.6   Let f be a function from the set A to the set B, and suppose that S and T are subsets of A.

(a) Show that $f[S \cup T] = f[S] \cup f[T]$.   (b) Show that $f[S \cap T] \subseteq f[S] \cap f[T]$.

Find a counterexample showing that $f[S] \cap f[T] \subseteq f[S \cap T]$ is not always true.

7.7   Are $f(x) = 4x + 1$ and $g(x) = 2(2x) + 1$ the same function? Discuss whether it is reasonable/unreasonable for them to be considered the same.

7.8   For each of the following functions, decide whether the function is *one-to-one, onto, both,* or *neither.*

(a) $f : \{1, 2, 3\} \to \{1, 2, 3, 4\}$ such that $f(x) = x + 1$.

(b) The function $\{ \langle a, 1 \rangle, \langle b, 2 \rangle, \langle c, 3 \rangle \}$ from $\{a, b, c\}$ to $\{1, 2, 3, 4\}$.

(c) The function $\{ \langle a, 1 \rangle, \langle b, 2 \rangle, \langle c, 3 \rangle, \langle d, 3 \rangle \}$ from $\{a, b, c, d\}$ to $\{1, 2, 3, 4\}$.

(d) The function $\{ \langle a, 4 \rangle, \langle b, 2 \rangle, \langle c, 3 \rangle, \langle d, 1 \rangle \}$ from $\{a, b, c, d\}$ to $\{1, 2, 3, 4\}$.

(e) $f : \mathbb{R} \to \mathbb{R}$ such that $f(x) = 2x + 1$.

(f) $f : \mathbb{N} \to \mathbb{N}$ such that $f(x) = x + 2$.

(g) The function from English words to the set of natural numbers that for each word gives the number of letters in the word.

(h) The function from English words to the set of letters that for each word gives the first letter of the word.

7.9   Consider whether the following statements are true for all functions $f : A \to B$ and $g : B \to C$. If they are true, provide an explanation; if not, provide a counterexample.

(a) If f and g are both injective, then $g \circ f$ is injective.

(b) If f and g are both surjective, then $g \circ f$ is surjective.

(c) If f and g are both bijective, then $g \circ f$ is bijective.

(d) If f is injective and g is surjective, then $g \circ f$ is bijective.

(e) If $g \circ f$ is surjective, then f is surjective.

(f) If g is injective, $g \circ f$ then is injective.

(g) If g is not injective, $g \circ f$ then is not injective.

7.10   Let A be a finite set. Is it the case that an injective function from A to A is always surjective? Conversely, is it the case that a surjective function from A to A is always injective? Why does this not mean that all surjective functions from a set to itself are injective, and vice versa? Explain.

Chapter 8

A Little More Set Theory

■■□□□□■■■Ⅹ■■■□□□□□□■■■■

In this chapter you will learn a little more set theory. You will learn about the universal set, the complement of a set, power sets, and more about Venn diagrams. Finally, we will talk about infinity, cardinality, and the concepts of countability and uncountability.

Set Theory

Set theory is both a separate branch of and a *foundation* for mathematics. The German mathematician *Georg Cantor* (1845–1918) is considered the founder of modern set theory, and it was from him that we got the concept of uncountable sets. This chapter is a natural continuation of Chapters 1 and 7, because we are going to see more *operations* on sets, and we will use *functions* to analyze the size of sets.

Set Complement and the Universal Set

In propositional logic we can express the *negation* of propositions, and this is generally a powerful operation. For example, what is the negation of a proposition that says that an element is a member of a set? What is the set of elements that are *not* contained in a set? We know what the answer is if we have a set that we can relate to: the set $A \setminus B$ consists of all the elements of A that are not in B. But what if we don't have a set like A? What is the set of elements that are *not* contained in $\{1, 2, 3\}$? Is *everything* else going to belong there? In order to answer these questions, it is useful to assume that there is always an underlying *universal set*:

Definition 8.1. The universal set

We assume that there is an underlying **universal set**, depending on context, and that U stands for this set. If nothing else is specified, we assume that U stands for an *arbitrary* universal set.

What the universal set is varies from context to context. Sometimes it will be the set of natural numbers, while at other times it may be the set of propositional formulas.

© Springer Nature Switzerland AG 2021
R. Antonsen, *Logical Methods*, https://doi.org/10.1007/978-3-030-63777-4_9

In the definitions of operations on sets, the universal set has not been necessary until now. In the definition of intersection, for example, $(A \cap B)$ is defined from the sets A and B alone. But in order to define the *complement* of a set, it is necessary.

Definition 8.2. Complement

If M is a set and U is the universal set, the **complement** of M is the set of all elements of U that are *not* contained in M. The complement of M is written $\overline{M}$.

Notice that $\overline{M}$ equals $U \backslash M$, which we already know. When we write $\overline{M}$, the universal set must be known, or we must assume that there is an *arbitrary* universal set. We can also draw diagrams for the complements of sets, but then we have to draw the universal set. For if not, where should the diagram end? We do not need to draw the universal set for the other set operations. In the following figures we use a rectangle around the circles to represent the universal set:

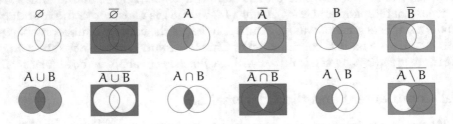

Q   **Complement.** Suppose that $U = \{1, 2, 3, 4, 5, 6\}$. Then we get

$$\overline{\varnothing} = \{1, 2, 3, 4, 5, 6\}, \qquad \overline{\{1, 2, 3\}} = \{4, 5, 6\}, \qquad \overline{\{1, 2, 3, 4, 5, 6\}} = \varnothing,$$
$$\overline{\{1\}} = \{2, 3, 4, 5, 6\}, \qquad \overline{\{1, 2, 3, 4\}} = \{5, 6\}, \qquad \overline{\{1, 3, 5\}} = \{2, 4, 6\},$$
$$\overline{\{1, 2\}} = \{3, 4, 5, 6\}, \qquad \overline{\{1, 2, 3, 4, 5\}} = \{6\}, \qquad \overline{\{2, 4, 6\}} = \{1, 3, 5\}. \qquad \blacklozenge$$

Q   **The complement and infinite sets.** Suppose that U is the set of natural numbers. The complement of $\{0, 1, 2, 3, 4\}$ is the set of all the natural numbers from 5 onward:

$$\overline{\{0, 1, 2, 3, 4\}} = \{5, 6, 7, 8, \ldots\}.$$

The complement of the set of even numbers is the set of odd numbers:

$$\overline{\{0, 2, 4, 6, \ldots\}} = \{1, 3, 5, 7, \ldots\}. \qquad \blacklozenge$$

Notice that $\overline{\varnothing} = U$ and $\overline{U} = \varnothing$, no matter what U is. We also have that $A \cup \overline{A} = U$ and $A \cap \overline{A} = \varnothing$. *Do you see the connection between this and propositional logic?*

Computing with Venn Diagrams

Suppose that $U = \{1, 2, 3, \ldots, 12\}$ and that the following sets are given:

$$A = \{x \mid x \text{ is odd}\} \qquad B = \{x \mid x > 7\} \qquad C = \{x \mid x \text{ is divisible by } 3\}$$

Now we can use Venn diagrams to calculate sets in a practical way:

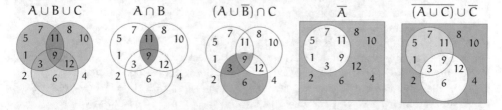

$A \cup B \cup C \qquad A \cap B \qquad (A \cup \overline{B}) \cap C \qquad \overline{A} \qquad \overline{(A \cup C)} \cup \overline{C}$

Venn Diagrams for Multiple Sets

Until now, we have only seen Venn diagrams for up to three sets, and a natural question is whether it is possible to use them for a larger number of sets. With two sets we get four different combinations in the Venn diagram ⊘. With three sets, we get eight different combinations in the Venn diagram ⊛. Note that each combination of sets is represented by exactly one shaded region in the diagram. For example, if the three sets are A, B, and C, and we insert the truth values for the three propositions $x \in A$, $x \in B$, and $x \in C$ in the different regions, we get exactly the same combinations as in a truth table. If we try to do this with four circles, as in diagram (1) below, we quickly discover that it is not possible to arrange four circles in such a way that all sixteen combinations of sets occur. In diagram (1) there are two combinations that are not represented by any region. *Try to identify which combinations these are.* But if we use curves other than circles, for example *ellipses*, it is possible to make Venn diagrams for more than three sets. Diagrams (2) and (3) represent correct Venn diagrams for four and five sets, respectively.

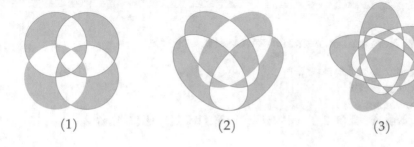

(1)   (2)   (3)

> **Digression**
>
> An n-Venn diagram is defined as a set of n curves in the plane such that each curve is closed and does not cross itself and such that all of the 2^n different combinations of sets are represented by one connected nonempty region. An n-Venn diagram is **symmetric** if it is identical after being rotated $360°/n$ around its center. In 1963, it was proved that symmetric n-Venn diagrams exist only when n is a prime number; that is, if an n-Venn diagram is symmetric, then n is a prime number. It was not until 2003 that someone was able to prove that symmetric Venn diagrams actually exist for all prime numbers. In other words, it is both a necessary and sufficient condition that n be a prime for a symmetric n-Venn diagram to exist. Furthermore, we say that a Venn diagram is **simple** if at most two lines intersect at a point. If we additionally require that the Venn diagrams be *simple*, the situation is much more difficult. In 1975, the Croatian-American mathematician *Branko Grünbaum* (1929–2018) discovered the simple symmetric 5-Venn diagram we have just seen, and in 1992, a simple symmetric 7-Venn diagram was discovered for the first time. For $n > 7$, it was an open problem until 2012, but then the mathematicians Mamakani and Ruskey found a simple symmetric 11-Venn diagram. No one knows whether a simple symmetric 13-Venn diagram exists.

Power Sets

Constructing *power sets* is yet another way to create new sets from old. When we form the power set of a set, we always get a set that is *larger* than the one we started with.

Definition 8.3. Power set

If M is a set, the **power set** of M is the set of all subsets of M. We write $\mathcal{P}(M)$ for the power set of M.

Q   Power set. Here are some simple power sets:

- The power set of $\varnothing$ is $\{\varnothing\}$.
- The power set of $\{1\}$ is $\{\varnothing, \{1\}\}$.
- The power set of $\{1, 2\}$ is $\{\varnothing, \{1\}, \{2\}, \{1, 2\}\}$.
- The power set of $\{1, 2, 3\}$ is $\{\varnothing, \{1\}, \{2\}, \{3\}, \{1, 2\}, \{1, 3\}, \{2, 3\}, \{1, 2, 3\}\}$.   ♦

If a finite set has n elements, the power set always has 2^n elements. That is because there are 2^n *subsets* of a set with n elements. One way to understand this is by observing that in forming all the subsets, there are two possibilities for each of the n elements: either it is an element of a subset or it is not an element of the subset.

Infinity

Imagine that you have a string of black and white pearls, stretching infinitely far in both directions. The pearls are strung such that two consecutive black pearls always alternate with one white pearl, in this way:

Are there more black than white pearls on the chain? It is tempting to answer yes, and indeed that there are twice as many, but we shall now see that it is equally meaningful to say that there are equally many of each type. We can compare the situation with sets of numbers: *which set is larger: the set of natural numbers or the set of even natural numbers?* Here it is also tempting to answer that there are more natural numbers, but for each of the natural numbers, there is *exactly* one even number, and vice versa, which should mean that there are equally many of each. We will now make this precise through the concept of *cardinality*.

Cardinality

Cardinality intuitively means "size" when we talk about sets. But what does it really mean that a set is *larger* than another set? For finite sets it is simple: a set is larger if it has more elements. But how about *infinite* sets? The following definition is now standard, but this has not always been the case.

Definition 8.4. Cardinality

Two sets M and N have the **same cardinality** if there is a one-to-one correspondence between the elements of M and N. We write $|M| = |N|$ when that is the case. The set M has **cardinality less than or equal** to that of N if there is a one-to-one correspondence between M and a subset of N. We write $|M| \leqslant |N|$ when that is the case. If M is a finite set, the cardinality of M equals the number of elements of M, and in that case we use the notation $|M|$ for the number of elements of M.

Q Same cardinality, finite sets. The definition of *same cardinality* works equally well for finite as for infinite sets. Let M be the set $\{1, 2, 3\}$ and let N be the set $\{a, b, c\}$. Then there is a bijection $f : M \to N$ such that $f(1) = a$, $f(2) = b$, and $f(3) = c$. The sets therefore have the same cardinality, and we can write $|M| = |N|$.

$$
\begin{array}{ccc}
1 & \longrightarrow & a \\
2 & \longrightarrow & b \\
3 & \longrightarrow & c
\end{array}
$$

♦

Q Cardinality, finite sets. The cardinality of the set $\{a, b, c\}$ equals 3, and we write $|\{a, b, c\}| = 3$. The cardinality of the empty set is equal to zero, so we write $|\varnothing| = 0$.♦

Q **Cardinality, infinite sets.** As usual, $\mathbb{N}$ denotes the set of natural numbers, $\{0, 1, 2, 3, 4, 5, \dots\}$. Let $2\mathbb{N}$ stand for the set of all even natural numbers, that is, $\{0, 2, 4, 6, 8, 10, \dots\}$. The function $f(x) = 2x$ from $\mathbb{N}$ to $2\mathbb{N}$ provides a one-to-one correspondence between $\mathbb{N}$ and $2\mathbb{N}$, and that means that $\mathbb{N}$ and $2\mathbb{N}$ have the same cardinality. We write $|\mathbb{N}| = |2\mathbb{N}|$. ◆

Countability

The concept of *cardinality* enables us to identify the following important property:

Definition 8.5. Countable

An infinite set M is **countable** if there is a one-to-one correspondence between the elements of M and the natural numbers. If not, M is **uncountable**. All finite sets are countable.

In other words, a set is countable if it has the same cardinality as a subset of the natural numbers. This means that all the elements of the set can be *counted* and that each element gets its own natural number. We have just seen that the set of even numbers is countable.

Q **Countability of integers.** The set of integers, $\mathbb{Z}$, is countable. In order to prove this, we must create a one-to-one correspondence between this set and the set of natural numbers. We can do this as follows:

x	0	1	2	3	4	5	6	7	8	9	10	11	12	13	14	$\cdots$
$f(x)$	0	−1	1	−2	2	−3	3	−4	4	−5	5	−6	6	−7	7	$\cdots$

All the natural numbers are listed in the top row, and all the integers are listed in the bottom row. For each natural number, there is exactly one integer, and vice versa. The function is *injective* because each integer appears exactly *once* in the bottom row, and it is *surjective* because *all* the integers occur in the bottom row. ◆

Q **Countability of fractions.** The set of fractions, $\mathbb{Q}$, that is, the set of numbers that can be written as m/n, where m and n are integers, and n does not equal 0, is countable. It may seem surprising, because between two fractions there is always another fraction. We may prove that this set is countable by organizing all the fractions in a big table. For convenience, we will only do it here for the positive fractions, but it is completely similar for the set of all fractions, including 0 and the negative ones.

In the table we see a method for listing all the fractions, by starting at the bottom left and systematically moving upward and to the right. This is a *surjective* function, because we have to come to every fraction sooner or later. In order to get a *one-to-one correspondence* with the natural numbers, we must skip all the fractions that we

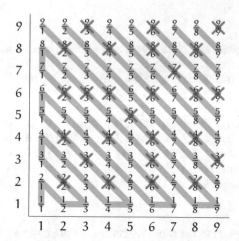

have already included. For example, when we get to $\frac{2}{2}, \frac{2}{4}, \frac{3}{3}, \frac{4}{2}, \frac{2}{6}, \frac{4}{4}, \frac{6}{2}$, etc., we skip them because we have already included them. In this way we get a list of all positive fractions, and it thus follows that the set is countable. ◆

Uncountability

There are sets that are so large that they are not countable, and Georg Cantor was the first to prove this. In an article from 1874 he gave a proof that the set of real numbers does not have the same cardinality as the natural numbers, in other words, that *the set of real numbers is not countable*.

In general, the power set of a set always has greater cardinality than the set itself. Power sets are also defined for infinite sets, and we can prove that there never exists a *surjective* function from a set to its power set. On the other hand, it is easy to define an *injective* function f, for example by $f(X) = \{X\}$.

Q **Uncountability via diagonalization.** Let us look at the set of all infinite sequences of the form

$$b_0 b_1 b_2 b_3 \ldots$$

where each b_i is either 0 or 1. We may think of such a sequence as a function from natural numbers to $\{0, 1\}$, which amounts to the same thing. Now we will prove that the set of all these sequences is *uncountable* with a proof by contradiction. Suppose for the sake of obtaining a contradiction that there is a one-to-one correspondence between this set and the natural numbers. Then we can make a list of all the elements like this:

```
sequence 0:  1  1  0  1  1  0  1  0  1  1  ···
sequence 1:  0  0  1  1  0  1  1  1  1  1  ···
sequence 2:  1  1  1  0  0  0  0  0  0  1  ···
sequence 3:  1  1  0  0  1  0  1  0  0  0  ···
sequence 4:  0  1  1  0  0  1  0  0  1  1  ···
sequence 5:  1  0  0  0  1  0  1  0  1  0  ···
sequence 6:  1  0  1  1  0  1  0  1  0  0  ···
sequence 7:  1  1  0  0  1  1  0  1  0  0  ···
sequence 8:  1  0  1  1  0  1  1  0  0  1  ···
sequence 9:  1  1  1  0  0  0  1  0  0  1  ···
```

By looking at the *diagonal* in the table we can now construct an infinite sequence that has to be different from each sequence in the list. We do so by letting the symbol in position n of the new sequence be the *opposite* of the symbol in position n of sequence n.

```
sequence x:  0  1  0  1  1  1  1  0  1  0  ···
```

This new sequence cannot be found anywhere in the list, because it is different from every sequence in at least one position. This argument holds in the same way for any attempt to make such a list. We can conclude that a one-to-one correspondence with the natural numbers does not exist and that the set of infinite sequences is not countable.   ♦

The method in this example can also be used to prove that the set of *real numbers* is uncountable. For any countable list of real numbers, we can construct a real number that is not included in the list. This is often called a **diagonalization argument**.

Digression

Ever since *Georg Cantor* discovered that uncountable sets exist, this topic has been discussed within both mathematics and philosophy. In 1878, Cantor proposed the **continuum hypothesis**, which says that there is no set with cardinality strictly between the cardinality of the natural numbers and that of the real numbers. One of the best-known *axiomatizations* of set theory, which took shape in the 1920s, is called **ZFC**, after the German mathematician *Ernst Zermelo* (1871–1953) and the Israeli mathematician *Abraham Fraenkel* (1891–1965). The Austrian logician and mathematician *Kurt Gödel* (1906–1978) proved in 1940 that the continuum hypothesis is consistent with ZFC, but in 1963 the American mathematician *Paul Cohen* (1934–2007) proved that the *negation* of the continuum hypothesis is also consistent with ZFC. This means that the continuum hypothesis is *independent* of the axioms in ZFC, and this is an important result in modern set theory.

Exercises

8.1 Suppose that $U = \{1, 2, 3, 4, 5, 6, 7, 8, 9\}$, and let $A = \{1, 2, 3, 4, 5, 6\}$, $B = \{1, 3, 5, 7\}$, and $C = \{2, 4, 6, 8\}$. Find the following sets.

(a) $\overline{A} \cup \overline{B}$ (b) $\overline{A} \cap \overline{B}$ (c) $\overline{B \cup C}$ (d) $\overline{B} \cap \overline{C}$ (e) $\overline{A \setminus B}$ (f) $\overline{A} \cup B$

8.2 Answer the following questions and justify your answers.

(a) What is the power set of $\varnothing$?
(b) What is the power set of $\{\varnothing\}$?
(c) What is the power set of $\{\pi, \varnothing\}$?
(d) What is the power set of the power set of $\varnothing$, that is $\mathcal{P}(\mathcal{P}(\varnothing))$?
(e) Is the claim "every set with n elements has 2^n subsets" true when $n = 0$?
(f) If a set X has n elements, how many elements does $\mathcal{P}(\mathcal{P}(X))$ have?

8.3 Suppose that A, B, and C are subsets of a universal set U. Show the following:

(a) $A \subseteq B$ if and only if $\overline{B} \subseteq \overline{A}$.
(b) $A \setminus B = A \cap \overline{B}$.
(c) $(A \cap B) \cup (A \cap \overline{B}) = A$.
(d) $A \cap \overline{(B \cup C)} = (A \setminus B) \cap (A \setminus C)$.
(e) $(\overline{A} \setminus B) \cap C = C \setminus (A \cup B)$.
(f) $\overline{A \cup B} = \overline{A} \cap \overline{B}$.

8.4 Let A and B be two sets.

(a) If $A \subseteq B$ is true, is it then the case that $\mathcal{P}(A) \subseteq \mathcal{P}(B)$?
(b) If $\mathcal{P}(A) \subseteq \mathcal{P}(B)$ is true, is it then the case that $A \subseteq B$?
(c) Is it possible that $A \subseteq \mathcal{P}(A)$?
(d) Is it necessary that $A \subseteq \mathcal{P}(A)$?
(e) Is it possible that $\mathcal{P}(A) \subseteq A$?
(f) What is the relationship between $\mathcal{P}(A \cup B)$ and $\mathcal{P}(A) \cup \mathcal{P}(B)$?
(g) What is the relationship between $\mathcal{P}(A \cap B)$ and $\mathcal{P}(A) \cap \mathcal{P}(B)$?

8.5 Show that $\overline{\overline{A} \cap B} = A \cup \overline{B}$ using Venn diagrams.

8.6 Determine which of the following sets are power sets of a set. If the set is a power set, find the underlying set; if not, explain why the set is *not* a power set.

(a) $\varnothing$ (c) $\{\varnothing, \{a\}\}$ (e) $\{\varnothing, \{a\}, \{\varnothing, a\}\}$
(b) $\{\varnothing\}$ (d) $\{\varnothing, \{1\}, \{1, 2\}\}$ (f) $\{\varnothing, \{\varnothing\}\}$

8.7 What is wrong with the following "Venn diagrams" for four sets?

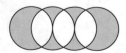

8.8 It is common to define the **symmetric difference** $A \triangle B$ between A and B as $(A \setminus B) \cup (B \setminus A)$.

  (a) Illustrate $A \triangle B$ with a Venn diagram.

  (b) Show that $(A \triangle B) \triangle C$ may be illustrated with the Venn diagram .

  (c) Why does this show that we can write $A \triangle B \triangle C$ without parentheses?

8.9 What is the cardinality of each of the following sets?

  (a) $\{a\}$ (c) $\{a, \{a\}, \{a, b, c\}\}$

  (b) $\{a, \{a\}\}$ (d) $\{a, \{a\}, \{a, \{a\}\}, \{\{a\}\}\}$

8.10 Suppose that $|S| = 3$ and $|T| = 4$, and that $S \cap T = \varnothing$.

  (a) What is $|S \times T|$?   (d) What is $|S \setminus T|$?

  (b) What is $|S \cup T|$?   (e) What is $|T \setminus S|$?

  (c) What is $|S \cap T|$?   (f) What are $|\mathcal{P}(S)|$ and $|\mathcal{P}(T)|$?

  Which of the answers depend on the assumption that $S \cap T = \varnothing$?

8.11 Prove that the following sets are countable. Hint: Find a one-to-one correspondence between each set and the natural numbers.

  (a) The set of positive odd numbers.

  (b) The set of negative integers.

  (c) The set of integers.

  (d) The set of all finite sequences of the form $b_1 b_2 \ldots b_n$, where $n \geqslant 1$ and each b_i equals either 0 or 1.

8.12 For each of the following properties, find examples of countable infinite sets S and T such that the property holds.

  (a) $S \setminus T$ is finite.   (b) $S \setminus T$ is infinite.   (c) $|S \setminus T| = 8$.

8.13 Show that if S and T are countable and $S \subseteq T$, then S is countable.

8.14 If A is uncountable and B is countable, is it the case that $A \setminus B$ must be uncountable? If yes, provide a proof; if no, explain why.

8.15 Prove that $|X| < |\mathcal{P}(X)|$, that is, the cardinality of $\mathcal{P}(X)$ is strictly greater than the cardinality of X. Use a diagonalization argument.

Chapter 9

Closures and Inductively Defined Sets

■·■·■·■·■·■·Ⓘ·■·■·■·■·■·■·■·■

In this chapter you will learn about closures, both of sets and of relations, and how sets can be constructed inductively. We will look at how several sets are defined inductively in this way: sets of numbers, bit strings, propositional formulas, lists, binary trees, and programming languages. In addition, we will become familiar with alphabets and strings, and how formal languages can be defined inductively.

Defining Sets Step by Step

Defining sets *inductively* is a way of defining sets that is useful and widely used in both computer science and mathematics. To define sets inductively means to build them up from *below* or *within*, and it is a way to construct sets *step by step*. Once we have defined a set inductively, we have good control of what items are contained in the set. It becomes easier to treat and prove statements about a set than if we defined it in a more abstract way. We will also see that to define a set inductively is to capture something complex using simple rules. As an important preparation and warm-up, we first look at *closures* of sets and relations.

Closures of Sets

An important property of inductively defined sets is that they are *closed* under given operations, and therefore we first look at the notion of closure.

Definition 9.1. Closed set

That a set is **closed** under a given operation means that when we perform that operation on one or more elements of the set, we are guaranteed that the result is an element of the same set. If M is a set, the smallest set containing M that is closed under one or more operations is called the **closure** of M under those operations.

© Springer Nature Switzerland AG 2021
R. Antonsen, *Logical Methods*, https://doi.org/10.1007/978-3-030-63777-4_10

Notice that there is nothing in this definition about what the domain of the operation is, and this is intentional. Normally, the domain consists of another set, one larger than the one being constructed. For if the domain of the operation *equals* the set that is being defined, it follows directly from the definition of an operation that the set is already closed. Here we are concerned primarily with the process of adding elements and how sets are constructed step by step. The definition of closure says that the closure of a set is the *smallest* containing set closed under the operation, and that means that it is a subset of all other sets with the same property. Such a set may not exist, and in that case, the closure of a set is undefined.

Q **Closure of sets of numbers.** The set of *natural numbers* is closed under addition and multiplication. If we take two natural numbers and multiply or add them, we always get another natural number. But the set is *not* closed under subtraction, because both 2 and 3 are natural numbers, but $2 - 3$ is not. The set of *rational numbers* is also closed under addition and multiplication. If we take two rational numbers $\frac{a}{b}$ and $\frac{c}{d}$ and multiply or add them, we get $\frac{ac}{bd}$ and $\frac{ad+bc}{bd}$, respectively, both of which are also rational numbers. But the set is *not* closed under taking square roots, because 2 is a rational number, but $\sqrt{2}$ is not.   ◆

Closures of Binary Relations

It sometimes happens that a binary relation on a set lacks some elements that would make it reflexive, transitive, or symmetric. In such situations it can be useful to be able to add those elements.

> **Definition 9.2. Closure of a binary relation**
>
> If R is a relation, the smallest relation containing R and having a given property is the **closure** of R with respect to this property. If the property at hand is reflexivity, symmetry, or transitivity, the closure is called respectively the **reflexive**, **symmetric**, or **transitive** closure of R.

By definition, the closure is the *smallest* relation, which means that it is a subset of all the other relations with the same property. When we say the "reflexive and transitive closure," we means the smallest relation with both of these properties.

Q **Closures of binary relations.** Let A be the set $\{1, 2, 3\}$.
Let the following relation be given: $\{\langle 1, 2\rangle\}$
reflexive closure: $\{\langle 1, 2\rangle, \langle 1, 1\rangle, \langle 2, 2\rangle, \langle 3, 3\rangle\}$
Symmetric closure: $\{\langle 1, 2\rangle, \langle 2, 1\rangle\}$
Transitive closure: $\{\langle 1, 2\rangle\}$
Symmetric and *transitive* closure: $\{\langle 1, 2\rangle, \langle 2, 1\rangle, \langle 1, 1\rangle, \langle 2, 2\rangle\}$

The elements $\langle 1, 1\rangle$ and $\langle 2, 2\rangle$ must be included here because both $\langle 2, 1\rangle$ and $\langle 1, 2\rangle$ are included   ◆

Q **Closures of binary relations.** Let A be the set $\{1, 2, 3\}$.
Let R be the following relation: $\{\langle 1, 1\rangle, \langle 1, 2\rangle, \langle 2, 3\rangle\}$
Reflexive closure: $R \cup \{\langle 2, 2\rangle, \langle 3, 3\rangle\}$
Symmetric closure: $R \cup \{\langle 2, 1\rangle, \langle 3, 2\rangle\}$
Transitive closure: $R \cup \{\langle 1, 3\rangle\}$
Reflexive and transitive closure: $R \cup \{\langle 1, 3\rangle, \langle 2, 2\rangle, \langle 3, 3\rangle\}$

Notice that R itself in included in all of the closures.   ♦

When we take the closure of a relation, we have to make sure that the relation in fact obtains the desired property. It is not enough to add only those tuples that are missing *at the outset*. Once those elements are added we have to continue and continue until we are finished. The next example shows this.

Q **Transitive closure.** What is the transitive closure of $\{\langle 1, 2\rangle, \langle 2, 3\rangle, \langle 3, 4\rangle\}$? We see that $\langle 1, 3\rangle$ and $\langle 2, 4\rangle$ are missing, but it is not sufficient to add only these, yielding the result $\{\langle 1, 2\rangle, \langle 2, 3\rangle, \langle 3, 4\rangle, \langle 1, 3\rangle, \langle 2, 4\rangle\}$, which is *not* a transitive relation. For the relation to become transitive, we must also add $\langle 1, 4\rangle$.   ♦

Inductively Defined Sets

We may think of the natural numbers either as the closure of the set $\{0\}$ under the operation that adds the number 1 or as a process whereby we begin with the number 0 and add new elements, one by one, infinitely often.

Definition 9.3. Inductively defined set

An **inductively defined set** is the smallest set that contains a given set – called the **base set**, or initial set – and is closed under given operations. A set is inductively defined in the following three steps:

 – The *base case*, or basis: to specify a base set.
 – The *inductive step*: to specify operations.
 – *The closure*: to construct the smallest set that includes the base set and is closed under the operations.

We may imagine the inductively defined set as a sort of onion with layer upon layer, one outside the other. At the core of the onion we have the base set, and for each of the layers, as we move outward, we add new elements. What we get in the end, after having added all possible elements, and not having added anything else, is the inductively defined set. We will use the rest of this chapter to inductively define several different sets that we will continue to work with in the following chapters.

Digression

The concept of a *fixed point* is important in computer science and mathematics. The definition is simple: If f is a function and $f(x) = x$, then x is called a **fixed point** of f. When we take the closure of a set under an operation, we obtain a *fixed point* (of the closure operation), because we don't get anything more by applying the operation more times on any of the elements. In mathematics and computer science there are many forms of *fixed-point theorems*, and these are usually theorems saying that a function has a fixed point under certain conditions. One of the best-known fixed-point theorems is *Brouwer's fixed-point theorem*, named after the Dutch mathematician and philosopher *L.E.J. Brouwer* (1881–1966). This says, much simplified, that if you stir a cup of coffee, there will always be at least one point that ends up at its initial position. In mathematics and computer science the concept of the *least fixed point* of a function is important. This concept can be used, for instance, to provide semantics for programming languages. In practice, it may be useful to look at the sequence $x, f(x), f(f(x)), f(f(f(x))), \ldots$ and how it behaves for a given element x, whether, for example, it approaches a fixed point for f. After reading about idempotency in Chapter 20, you will see, for example, that a unary operation f is idempotent if it maps all points x to a fixed point for f; that is, $f(f(x)) = f(x)$.

Sets of Numbers

Q **Natural numbers, inductively defined.** The set of natural numbers can be inductively defined as the smallest set $\mathbb{N}$ such that

 – $0 \in \mathbb{N}$ and
 – if $x \in \mathbb{N}$, then $x + 1 \in \mathbb{N}$.

Here, the base set is $\{0\}$, and $\mathbb{N}$ is closed under the successor function. There are many other sets that include the base set and are closed under the successor function, for example $\{0, \frac{1}{2}, 1, \frac{3}{2}, \ldots\}$ and the real numbers, but we are looking for the *smallest* of these.   ◆

Q **Even numbers, inductively defined.** The set of even numbers is the smallest set that contains 0 and is closed under the operations that add and subtract two, respectively. Here, the base set is $\{0\}$ and the operations are:

 – one that takes x as an argument and returns $x + 2$,
 – one that takes x as an argument and returns $x - 2$.

The elements of the base set:	0
New elements in step 1:	2, −2
New elements in step 2:	4, −4
New elements in step 3:	6, −6

There are infinitely many other sets containing 0 that are closed under exactly these operations. Both the integers and the real numbers are. But the set we are defining is the *smallest* of all of these.   ◆

Propositional Formulas

Now we can return to propositional logic and look at the definition of the set of propositional formulas. This was an inductive definition.

Definition 9.4. Propositional formula

The set of **propositional formulas** is the smallest set X such that the following holds:

- Each propositional variable is in X. These make up the base set, whose elements are called **atomic formulas**.
- If F is in X, then ¬F is in X.
- If F and G are in X, then $(F \land G)$, $(F \lor G)$, and $(F \to G)$ are in X.

The set of propositional formulas is the smallest set that includes all the propositional variables and is closed under the operation of combining formulas with connectives. Each propositional formula is therefore either a propositional variable or a composition of propositional formulas, that is, of the form ¬F, $(F \land G)$, $(F \lor G)$, or $(F \to G)$.

Lists and Binary Trees

Most programming languages define a type of *list*. A list is a finite sequence of objects, and from a mathematical point of view, lists are not very different from *tuples*. But when we talk about lists, they usually come with a lot of terminology and structure. For example, it is common that the elements of a list must be read, or *accessed*, according to how they are added to the list. First, we introduce some notation, and then we define the set of lists inductively.

Notation. We denote lists by providing the elements of the list separated by commas and contained inside a pair of parentheses (and). The function :: takes two arguments, an element x and a list L, and returns the list obtained by appending the element x to the beginning of L. For example, $(1 :: (2,3,4)) = (1,2,3,4)$, and $(4 :: ()) = (4)$. As long as it is clear, we may drop parentheses around $(x :: L)$ and write $x :: L$, but the parentheses that indicate the list must be in place.

Definition 9.5. List

The set of **lists** over a set A is inductively defined as the smallest set such that the following holds:

- () is a list over A. This is the **empty list** over A.
- If $x \in A$ and L is a list over A, then $x :: L$ is a list over A.

Q **Lists.** In order to get the list $(1, 2, 3)$ over the natural numbers $\mathbb{N}$ we may begin with the empty list and add elements from $\mathbb{N}$ using $::$ in the following way:

$$
\begin{aligned}
1 &:: (2 :: (3 :: ())) \\
&= 1 :: (2 :: (3)) \\
&= 1 :: (2, 3) \\
&= (1, 2, 3).
\end{aligned}
$$
◆

Another type of structure commonly used, particularly in computer science, is the *binary tree*. Trees are important because they enable us to create and represent information in a way that is a not *one-dimensional* and has additional structure. In Chapter 22, when we learn about graph theory, we shall define trees in a more abstract way. Here, the intuition is that the smallest tree is the empty tree, and that if L (*left*) and R (*right*) are two trees, then (L, x, R) is also a tree.

Definition 9.6. Binary tree

The set of **binary trees** over a set A is the smallest set such that the following hold:

- $()$ is a binary tree over A. This is called the **empty binary tree** or just the **empty tree**.
- If $x \in A$, and L and R are binary trees over A, then (L, x, R) is a binary tree over A. The element x is called a **node** in the binary tree. When a binary tree is of the form $((\,), x, (\,))$, then x is called a **leaf node**.

As long as it is unambiguous, we often refer simply to *trees* even when we are talking about *binary trees*.

Q **Binary trees.** Here are three trees over the natural numbers along with a figure of each tree. It is common not to draw the empty trees, but in these figures they are included. Trees are also sometimes drawn upside down.

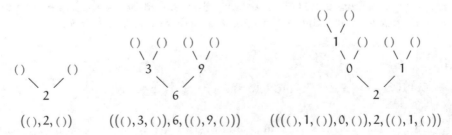

$$((\,), 2, (\,)) \qquad (((\,), 3, (\,)), 6, ((\,), 9, (\,))) \qquad ((((\,), 1, (\,)), 0, (\,)), 2, ((\,), 1, (\,)))$$

We can also draw the same trees more compactly, as in •, ⋎, and ⋎, with the empty trees not shown and the elements replaced with dots. ◆

Programming Languages

Programming languages are almost always defined inductively. The base set may consist of the simplest programs, for example assignments of values to variables, and the operations may be ways to compose smaller programs into larger ones. The inductively defined set we eventually get is the set of syntactically correct programs in the programming language. A program that checks whether another program is correctly written uses the inductive structure to check whether the program is built up from the base set and the permissible operations. For this check to work well, it must be the *smallest* set of programs we are checking against.

Alphabets, Characters, Strings, and Formal Languages

Formal languages can be defined as inductively defined sets. In Chapter 23 we will look at other methods for defining formal languages, but here we will use what we already know and get to know some concepts by defining sets inductively. Briefly, a *string* is a finite sequence of *characters*, taken from an underlying *alphabet*, and a *language* is a subset of all possible strings. Just as there is no effect on a number by adding zero or to a set by taking its union with the empty set, it is useful to have a string that has no effect when we combine it with another string. We call this the *empty string*. The following definition makes all this precise.

Definition 9.7. Alphabet, character, string

Suppose that A is a set of characters. The set A is called an **alphabet**. The set of **strings** over A is the smallest set A^* such that:

- $\Lambda \in A^*$, where the symbol Λ stands for **the empty string**;
- if $s \in A^*$ and $x \in A$, then $sx \in A^*$, where sx stands for the result of appending x to s. We write s instead of Λs.

A **language** over A is a subset of A^*.

This definition is quite compact, and some comments are called for. The alphabet, the set A, may be anything, but it is common to use an appropriate set of symbols. Our symbol for the empty the string is Λ, the uppercase Greek letter *lambda*, but it is also common to use other symbols, like λ, the lowercase Greek letter *lambda*; ϵ or ε, the lowercase Greek letter *epsilon*; or e. Notice that the definition is inductive: The base set consists of the empty string, and at each inductive step, strings are created that are one character longer than those we had before. Also note the use of the star symbol. If A is an alphabet, then A^* is a language over this alphabet. If A contains at least one character, then A^* contains strings that are arbitrarily, but not infinitely, long.

Q Sets of strings. Here are the sets of strings over some simple alphabets:

$$\varnothing^* = \{\Lambda\},$$
$$\{a\}^* = \{\Lambda, a, aa, aaa, aaaa, aaaaa, \ldots\},$$
$$\{0, 1\}^* = \{\Lambda, 0, 1, 00, 01, 10, 11, 000, 001, \ldots\},$$
$$\{a, b, c\}^* = \{\Lambda, a, b, c, aa, ab, ac, ba, bb, bc, ca, cb, cc, aaa, \ldots\}. \qquad \blacklozenge$$

How do we specify a language? We are free to do it however we want, as long as we do it precisely, but it is useful to define languages as inductively defined sets.

Q All strings with an even number of characters. Let S be the smallest set such that $\Lambda \in S$, and if $t \in S$, then taa, tab, tba, tbb $\in S$. We get that aaaa $\in S$, but aaa $\notin S$. The set S contains all strings over $\{a, b\}$ that have an even number of characters. $\blacklozenge$

Definition 9.8. Concatenation

The operation consisting of joining two strings s and t together into one, st, is called **concatenation**. The notation s^n is used as an abbreviation for the string s repeated n times; for example, $a^3 b^3$ is the same as aaabbb. We assume that $s^0 = \Lambda$.

Note. We allow ourselves here to disregard all the instances of the empty string, and the result of concatenating a string s with the empty the string is s, that is, $\Lambda s = s\Lambda = s$.

Q The language $a^n b a^n$. One language over the alphabet $\{a, b\}$ is the set of all strings of the form $a^n b a^n$, that is, the set $\{b, aba, aabaa, \ldots\}$. We can describe this language as $\{a^n b a^n \mid n = 0, 1, 2, \ldots\}$. This language can be defined inductively as the smallest set S such that $b \in S$, and if $t \in S$, then ata $\in S$. $\blacklozenge$

Q The set of symmetric strings. The set S of symmetric strings with an even number of characters over the alphabet $\{a, b\}$ can be defined inductively as the smallest set S such that $\Lambda \in S$, and if $s \in S$, then asa $\in S$ and bsb $\in S$. Here we see the first elements of S:

Elements in the base set:	Λ		
New elements in step 1:	aa	bb	
New elements in step 2:	aaaa	abba	baab	bbbb
New elements in step 3:	aaaaaa	aabbaa	abaaba	abbbba
	baaaab	babbab	bbaabb	bbbbbb

$\blacklozenge$

Q Languages we already know. The set of propositional formulas is a language over the alphabet $\{P, Q, R, \ldots, \wedge, \vee, \rightarrow, \neg, (,)\}$; the set of finite sequences of 0's and 1's is a language over the alphabet $\{0, 1\}$; and the set of syntactically correct programs in the programming language Java is a language. $\blacklozenge$

Bit Strings

A *bit string* can be defined as a finite sequence of 0's and 1's, but by defining bit strings *inductively*, we get more control.

Definition 9.9. Bit string

The set of **bit strings** is inductively defined as the smallest set such that 0 and 1 are bit strings, and if b is a bit string, then b0 and b1 are also bit strings.

In the definition of bit strings the base set is $\{0, 1\}$, and the operations are:

- one that takes b as argument and returns b0;
- one that takes b as argument and returns b1.

Elements in the base set: 0   1
New elements in step 1:   00   10   01   11
New elements in step 2:   000   100   010   110   001   101   011   111

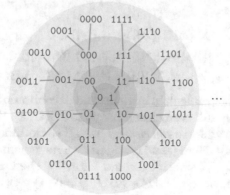

The set of bit strings as an inductively defined set.

We get the set of all bit strings by continuing this process to infinity. The set thus defined is infinite, but each individual bit string of the set has *finitely many* characters and occurs after *finitely many* steps. Notice that the set is completely unambiguously defined. We know exactly what is in the set and what is not in the set. Nothing but what is added with the help of the two operations is included. When we define a set inductively, we define it *from within* and in a constructive way. It gives us more information to do it in this way than, for example, by defining bit strings as sequences of 0's and 1's. Now we can, for example, find the length of a bit string by checking in which "layer" it was created.

Two Interesting Constructions

We finish the chapter with two sets that have a rich and interesting story.

Q   **The Thue–Morse sequence.** We shall now define a language over the alphabet {□, ■}. If X is a string over this alphabet, let $\overline{X}$ stand for the result of replacing all occurrences of □ with ■ and vice versa. For example, $\overline{□}$ = ■ and $\overline{■□}$ = □■. Let **T** be the smallest set such that □ ∈ **T**, and if X ∈ **T**, then X$\overline{X}$ ∈ **T**. Here are the smallest elements of **T** after □:

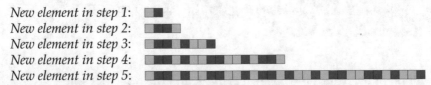

New element in step 1:
New element in step 2:
New element in step 3:
New element in step 4:
New element in step 5:

At each step, the next string is twice as long as the previous one. The infinitely long sequence the strings are approaching is called the **Thue–Morse sequence**, named after the Norwegian mathematician *Axel Thue* (1863–1922) and the American mathematician *Marston Morse* (1892–1977). It has many remarkable features. Among other things, there is no part of it that is of the form XXX, where X ∈ {□, ■}*.   ◆

Q   **Numbers as sets.** Let $\mathcal{O}$ be the smallest set such that ∅ ∈ $\mathcal{O}$ and if X ∈ $\mathcal{O}$, then X ∪ {X} ∈ $\mathcal{O}$. Here are the smallest, nonempty elements of $\mathcal{O}$:

$$\varnothing \cup \{\varnothing\} = \{\varnothing\} \in \mathcal{O},$$
$$\{\varnothing\} \cup \{\{\varnothing\}\} = \{\varnothing, \{\varnothing\}\} \in \mathcal{O},$$
$$\{\varnothing, \{\varnothing\}\} \cup \{\{\varnothing, \{\varnothing\}\}\} = \{\varnothing, \{\varnothing\}, \{\varnothing, \{\varnothing\}\}\} \in \mathcal{O},$$
$$\{\varnothing, \{\varnothing\}, \{\varnothing, \{\varnothing\}\}\} \cup \{\{\varnothing, \{\varnothing\}, \{\varnothing, \{\varnothing\}\}\}\} = \{\varnothing, \{\varnothing\}, \{\varnothing, \{\varnothing\}\}, \{\varnothing, \{\varnothing\}, \{\varnothing, \{\varnothing\}\}\}\} \in \mathcal{O}.$$

But this quickly becomes difficult to read! Instead of continuing like this, let us think about what is going on. If X is an element of $\mathcal{O}$, so is X ∪ {X}. *What does that mean?* It means that X must be added as *an element* to X. If something of the form {●, ●} is there, so is {●, ●, {●, ●}}. Here {●, ●} is added as a new element. Let us introduce some abbreviations: Let 0 stand for ∅ and 1 for {0}. Then the first line says that 1 ∈ $\mathcal{O}$. In the next step, we get 1 ∪ {1} = {0} ∪ {1} = {0, 1} ∈ $\mathcal{O}$, and it is natural to let 2 stand for {0, 1}. Now there is a pattern. Each time we get a new element of $\mathcal{O}$, this is the set of all the previous elements:

$$\varnothing \cup \{0\} = \{0\} = 1 \in \mathcal{O}, \qquad \{0, 1\} \cup \{2\} = \{0, 1, 2\} = 3 \in \mathcal{O},$$
$$\{0\} \cup \{1\} = \{0, 1\} = 2 \in \mathcal{O}, \qquad \{0, 1, 2\} \cup \{3\} = \{0, 1, 2, 3\} = 4 \in \mathcal{O}.$$

This is a set-theoretical way of defining the natural numbers that goes back to the Hungarian-American mathematician *John von Neumann* (1903–1957). By starting with the empty set, we can actually *define* the natural numbers in this way. Sets of this form are called **ordinal numbers**, and they are an important part of set theory. The next "natural" step in this construction is to take $\mathcal{O} \cup \{\mathcal{O}\}$, but we shall stop here.   ◆

Exercises

9.1 Let the following represent the relations R_1, R_2, and R_3 on the set $\{1, 2, 3, 4\}$:

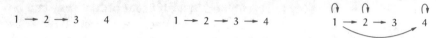

(a) For each relation, find its *reflexive* closure.
(b) For each relation, find its *symmetric* closure.
(c) For each relation, find its *transitive* closure.
(d) For each relation, find the smallest equivalence relation that extends it.

9.2 Suppose that $U = \{1, 2, 3, a, b\}$ and let the relation R on U be given by $R = \{\langle 2, 3 \rangle, \langle 3, 2 \rangle, \langle 1, a \rangle\}$.

(a) What is the reflexive closure of R?
(b) What is the symmetric closure of R?
(c) What is the transitive closure of R?

9.3 (a) Let R be a relation on a set S, and let Δ be the set $\{\langle x, x \rangle \mid x \in S\}$. Is it true that the reflexive closure of R is identical with $R \cup \Delta$? Explain.
(b) Let R be a relation on a set S, and let Σ be the set $\{\langle x, y \rangle, \langle y, x \rangle \mid x, y \in S\}$. Is it true that the symmetric closure of R is identical to $R \cup \Sigma$? Explain.
(c) Suppose that you have a relation R on a set S. If you first take the symmetric closure and then take the transitive closure, do you get a reflexive relation?
(d) Suppose that you have a relation R on a set S. If you first take the symmetric closure and then take the transitive closure, do you get the same result by doing things in the reverse order, that is, by first taking the transitive closure and then the symmetric closure?

9.4 Is it possible to take the *irreflexive* closure of a relation? If so, when is it possible?

9.5 Let R be a relation on a set S, and let R^* be the transitive closure of R. Show how R^* can be defined *inductively*.

9.6 Normally we cannot "see into" sets by means of $\in$. For example, $2 \notin \{1, \{2\}\}$. Explain how a relation $\in^*$ can be defined such that this becomes possible, for example, such that $2 \in^* \{1, \{2\}\}$ and $2 \in^* \{\{\{\{2\}\}\}\}$. Do the same with $\subseteq$.

9.7 In the inductive definition of binary trees over a set A, the smallest tree is defined as the *empty tree*. What if we changed the base case and defined the smallest trees directly from A? For example, we could say that if x is an element of A, then x, or $\langle x \rangle$, is a tree. Are we allowed to do this? What are the advantages and disadvantages of such a definition?

9.8   Can we have trees whose nodes are themselves trees? Can we have lists whose elements are themselves lists?

9.9   There is a difference between the trees $((( ), b, ( )), a, ( ))$ and $(( ), a, (( ), b, ( )))$. They have identical subtrees, but they are combined in a different order. They can be drawn as ⬩ and ⬩, respectively. The difference arises because all trees have a left side and a right side. Define binary trees inductively in such a way that this order does not matter, in other words, so that we could as well have drawn ⬩.

9.10   In the definition of binary trees, two trees are combined to create new trees, and the branching is binary. Find another way to define trees inductively such that more trees can be combined at each step and such that we can have more branching at each node:

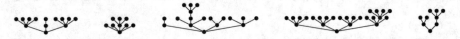

9.11   (a) Give an inductive definition of the language $\{a^n, b^n \mid n = 0, 1, 2, \dots\}$, that is, $\{\Lambda, a, b, aa, bb, aaa, bbb, \dots\}$.

   (b) Give an inductive definition of the language $\{a^n b^n \mid n = 0, 1, 2, \dots\}$, that is, $\{\Lambda, ab, aabb, aaabbb, \dots\}$.

   (c) Give an inductive definition of the language $\{(ab)^n \mid n = 0, 1, 2, \dots\}$, that is, $\{\Lambda, ab, abab, ababab, \dots\}$.

9.12   For each of the following inductive definitions, start with the base set and construct the first ten elements of the set.

   (a) *Base set*: $\{3\}$. *Inductive step*: If x is an element, then $2x - 1$ is an element.

   (b) *Base set*: $\{1\}$. *Inductive step*: If x is an element, then $2x$ and $2x + 1$ are elements.

   (c) *Base set*: $\{1\}$. *Inductive step*: If x is an element, then $3x$ and $3x + 1$ are elements.

   (d) *Base set*: $\{5\}$. *Inductive step*: If x is an element, then $10x$ is an element.

   (e) *Base set*: $\{\varnothing\}$. *Inductive step*: If X is an element, then $\{X\}$ is an element.

   (f) *Base set*: $\{\Lambda, a, b\}$. *Inductive step*: If x is an element, then axa and bxb are elements.

9.13   Find inductive definitions for the following sets. If, for example, $\{1, 3, 5, 7, 9, \dots\}$ is the set, it can be inductively defined as the smallest S such that $1 \in S$, and if $x \in S$, then $x + 2 \in S$.

   (a) $\{2, 4, 6, 8, 10, 12, \dots\}$
   (b) $\{1, 4, 7, 10, 13, 16, \dots\}$
   (c) $\{1, 2, 4, 8, 16, 32, 64, \dots\}$
   (d) $\{1, 3, 7, 15, 31, \dots\}$
   (e) $\{-5, -3, -1, 1, 3, 5, \dots\}$

   (f) $\{\dots, -7, -4, -1, 2, 5, 8, \dots\}$
   (g) $\{4, 7, 10, 13, \dots\} \cup \{3, 6, 9, 12, \dots\}$
   (h) $\{3, 16, 29, 42, \dots\}$
   (i) $\{3, 4, 5, 8, 9, 12, 16, 17, 20, 24, 33, \dots\}$
   (j) $\{1, 4, 9, 16, 25, \dots\}$

Chapter 10

Recursively Defined Functions

In this chapter you will learn to define functions recursively and how such functions are based on inductively defined sets. We will consider recursively defined functions on a series of sets: sets of numbers, bit strings, propositional formulas, lists, binary trees, and formal languages.

A Powerful Tool

Recursively defined functions are some of the most powerful tools in mathematics and computer science. We begin with inductively defined sets, which we studied in the previous chapter, and define functions recursively from these. The intuition is that we define a function recursively by first providing values for all the elements of the base set and then providing values for all the other elements by following the structure of the inductively defined set. It is therefore the case that functions are defined recursively in somewhat the same way as sets are defined inductively. But before we get to the actual definition, let us see an example of a recursively defined function and speak a little bit about the terminology.

The Triangular Numbers

We can see a recursively defined function by looking at the **triangular numbers**. These are the numbers of objects that can be arranged in equilateral triangles in the following way:

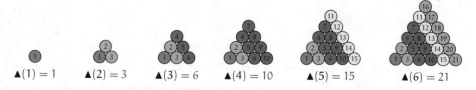

$$\blacktriangle(1) = 1 \quad \blacktriangle(2) = 3 \quad \blacktriangle(3) = 6 \quad \blacktriangle(4) = 10 \quad \blacktriangle(5) = 15 \quad \blacktriangle(6) = 21$$

We see that there is a connection between the *previous* triangular number and the *next*. If we let $\blacktriangle(n)$ stand for the nth triangular number, we get that $\blacktriangle(4) = \blacktriangle(3) + 4 = 6 + 4 = 10$. If we generalize over this, we get that $\blacktriangle(n + 1) = \blacktriangle(n) + (n + 1)$, and this is an example of a recursively defined function.

© Springer Nature Switzerland AG 2021
R. Antonsen, *Logical Methods*, https://doi.org/10.1007/978-3-030-63777-4_11

Induction and Recursion

It is the existence of a set that is inductively defined that allows us to define a function recursively. In Chapters 11 and 12 we shall build on this theory and look at how proofs by induction can be used to prove statements about inductively defined sets and recursively defined functions. These three are closely related: inductively defined sets, recursively defined functions, and proofs by induction. But before we look at the details, we are going to look at two phenomena that are often only assumed without being explicitly stated: *placeholders* and *equality*.

Form, Content, and Placeholders

We use *placeholders* all the time. A placeholder is a word or a variable that can stand for something else. It is important to be able to identify what the placeholders are in expressions. Here are four different expressions, and while they all differ in their *form*, they have exactly the same *content*:

$$f(x) = 2x + 1, \qquad f(y) = 2y + 1, \qquad f(\square) = 2\square + 1, \qquad f(\text{input}) = 2\text{input} + 1.$$

All the expressions represent that f is a function that multiplies its argument by 2 and adds 1. Each of x, y, $\square$, and input acts as a placeholder. In exactly the same way, it does not matter whether we write A or $\square$ in the following expressions:

> For all formulas A, A is valid or $\neg$A is valid.
> For all formulas $\square$, $\square$ is valid or $\neg\square$ is valid.

It is important to be aware that in a given context, some symbols may be *reserved* and not used as placeholders. For example, no one would use the symbols $+$, $-$, $\cdot$, $/$ as placeholders in mathematical expressions. Imagine that someone wrote $f(-) = 2 - +1$. This is not only difficult to read, but in the worst case, incorrect. In many programming languages it is forbidden to use reserved words freely; they often have an intended meaning and should not be used for anything else, especially not as names of variables. It is up to us to use symbols that do not create confusion. Sometimes we want to indicate what kind of *form* an expression has, and then we use placeholders.

$$(1 + 4)/(2 + 3) \text{ is of the form } x/y,$$
$$17 = 16 + 1 \text{ is of the form } n + 1,$$
$$(P \to Q) \to (Q \to R) \text{ is of the form } F \to G,$$
$$aabb \text{ is of the form } sx.$$

We shall soon see that this notation is frequently used when we define functions recursively.

Replacing Equals by Equals

It is often implicitly understood that we can replace equals by equals. Suppose, for example, that $A = B$. In a large expression in which A occurs, we can replace A with B because they are equal. For example, $(2 + 3A)$ equals $(2 + 3B)$. Logical equivalences are often used in this way, whereby one subexpression is replaced with one that is equivalent. When we work with recursive definitions, we replace equals by equals all the time without it being explicitly mentioned.

Recursively Defined Functions

The word **recursion** comes from Latin and means to "run back" or "return." In mathematics it has a very special meaning.

Definition 10.1. Recursively defined function

If a set M is inductively defined, we can define a function f on the domain M **recursively** in the following way:

- For each element x in the base set of M, specify a value for $f(x)$. This is called the **base case** for the function.
- For each element x in M that appears in an inductive step, define the value of $f(x)$ using the previously defined values for f. This is called the **recursive step**.

When a set is inductively defined, we have good control over what elements are in the set, and this is what enables us to define functions recursively on the set. The recursive definition of a function follows the inductive definition precisely. Now we will see many examples of recursive definitions. We begin by looking at functions on numbers, and thereafter bit strings and formal languages.

Number Sets

Q **A simple function on the natural numbers.** We may define a function from the natural numbers to the natural numbers recursively in the following way: Let $d(0) = 0$. This defines the value of each element of the base set. Let $d(n+1) = d(n)+2$, for all $n \in \mathbb{N}$. This defines the values of the other natural numbers. Based on this, we can calculate the first values:

$$d(1) = d(0) + 2 = 0 + 2 = 2,$$
$$d(2) = d(1) + 2 = 2 + 2 = 4,$$
$$d(3) = d(2) + 2 = 4 + 2 = 6.$$

What is this function doing? Can you say in your own words what it does? We see that $d(n+1)$ is defined *by means* of $d(n)$, a previously defined value. This is characteristic of recursively defined functions. When the value of $d(n)$ is known, then the value of $d(n+1)$ is also known. When a function is defined in this way, it is the *form* of the argument, which in this case is a number, that determines what the value will be. Each number is either 0 or of the form $n+1$, and therefore it is clear what the value will be. If the argument is 15, it is of the *form* $14+1$, and then we get $d(15) = d(14+1) = d(14) + 2$. We can also compute "from outside and in" in the following way:

$$
\begin{aligned}
d(3) &= d(2) + 2 \\
&= (d(1) + 2) + 2 \\
&= ((d(0) + 2) + 2) + 2 \\
&= ((0 + 2) + 2) + 2 \\
&= (2 + 2) + 2 \\
&= 4 + 2 \\
&= 6.
\end{aligned}
$$

We see that the function doubles the value of the argument. ◆

Q **The factorial function, recursively defined.**   The factorial function is often defined by writing $n! = 1 \cdot 2 \cdot 3 \cdots n$, and in this case it is easy to understand what the dots mean. Such dots often represent a recursively defined function. Now we shall define the **factorial function** recursively:

– Let $0! = 1$.
– Let $(n+1)! = (n+1) \cdot n!$, for all $n \in \mathbb{N}$.

We calculate the first values of f by following the definition:

$$
\begin{aligned}
1! &= (0+1) \cdot 0! = 1 \cdot 1 = 1, \\
2! &= (1+1) \cdot 1! = 2 \cdot 1 = 2, \\
3! &= (2+1) \cdot 2! = 3 \cdot 2 = 6, \\
4! &= (3+1) \cdot 3! = 4 \cdot 6 = 24.
\end{aligned}
$$

It is the form of the argument of the function that determines the value. ◆

Q **The Fibonacci numbers.**   One of the most famous number sequences in history is the **Fibonacci numbers**, where each number is the sum of the two previous ones:

$$0, 1, 1, 2, 3, 5, 8, 13, 21, \ldots .$$

These can be defined recursively in the following way:

– $F(0) = 0$ and $F(1) = 1$,
– $F(n+2) = F(n) + F(n+1)$ for all natural numbers n.

Note that we use two previously defined values here, both $F(n)$ and $F(n+1)$, when we define the value of $F(n+2)$. ◆

Digression

The Fibonacci numbers are named after the Italian mathematician *Leonardo Pisano Fibonacci* (ca. 1170–1250), who described the numbers in his work *Liber Abaci* of 1202. These numbers show up in many places in mathematics, and they have many interesting properties. For example, if you divide each Fibonacci number by the previous one, you get a sequence of numbers that approaches the **golden ratio**, which is

$$\varphi = \frac{1 + \sqrt{5}}{2} = 1.6180339887498\ldots.$$

You actually get this number regardless of which two numbers you begin the sequence with. If you start with 2 and 1, you get the **Lucas numbers**, named after the French mathematician *Édouard Lucas* (1842–1891):

$$2, 1, 3, 4, 7, 11, 18, 29, 47, 76, 123, \ldots.$$

The golden ratio also has many interesting properties: Multiplying it by itself is the same as adding one, $\varphi^2 = \varphi + 1$, and taking the inverse is the same as subtracting one, $1/\varphi = \varphi - 1$. If we convert the golden ratio to an angle, the **golden angle**, approximately 137.507°, we get the figure on the right by placing small circles next to each other from a central point and moving through this angle around the center for each new circle we add.

Bit Strings

Q **The value of a bit string, recursively defined.** We can recursively define a function v from bit strings to the natural numbers in the following way: Let $v(0) = 0$ and $v(1) = 1$. Here all the elements of the base set get values. If b is a bit string, let $v(b0) = 2 \cdot v(b)$ and $v(b1) = 2 \cdot v(b) + 1$. Here all the other elements get their values. Notice that the definition follows the inductive definition of bit strings: If the argument of v ends with 0, apply v to what is in front of 0 and multiply the result by 2. If the argument of v ends with 1, apply v to what is in front of 1, multiply the result by 2, and add 1. We can investigate what the function does on the bit strings 100 and 101 in this way:

$$
\begin{aligned}
v(100) &= 2 \cdot v(10) \\
&= 2 \cdot (2 \cdot v(1)) \\
&= 2 \cdot (2 \cdot 1) \\
&= 2 \cdot 2 = 4,
\end{aligned}
\qquad
\begin{aligned}
v(101) &= 2 \cdot v(10) + 1 \\
&= 2 \cdot (2 \cdot v(1)) + 1 \\
&= 2 \cdot (2 \cdot 1) + 1 \\
&= 2 \cdot 2 + 1 = 4 + 1 = 5.
\end{aligned}
$$

♦

❷ What does this function do? Can you say in your own words what it does? Hint: In the figure below, the values of the function v are written under the arguments.   ◆

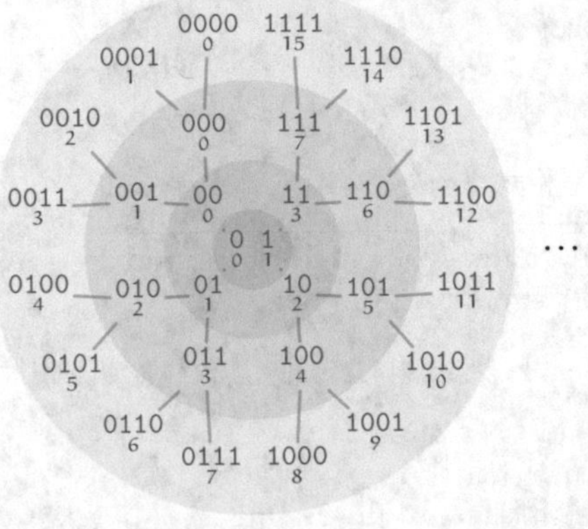

The values of bit strings.

Propositional Formulas

A valuation can be recursively defined as a function from propositional formulas to $\{0, 1\}$ in the following way.

Definition 10.2. Valuation

Let t be an assignment of truth values to propositional variables, that is, a function from the set of propositional variables to $\{0, 1\}$. A **valuation** can now be defined recursively as a function v from propositional formulas to $\{0, 1\}$ in the following way:

- $v(P) = 1$ if and only if $t(P) = 1$, for all propositional variables P.
  This is the *base case* of the definition.
- $v(\neg F) = 1$ if and only if $v(F) = 0$
- $v(F \wedge G) = 1$ if and only if $v(F) = 1$ and $v(G) = 1$
- $v(F \vee G) = 1$ if and only if $v(F) = 1$ or $v(G) = 1$
- $v(F \rightarrow G) = 1$ if and only if $v(F) = 1$ implies $v(G) = 1$.
  These four points make up the *recursive step* of the definition.

The last item of the definition could have been stated in the following way:

- $v(F \rightarrow G) = 1$ if and only if $v(F) = 0$ or $v(G) = 1$.

This is a typical example of a recursive definition: in the inductive definition of propositional formulas, the base set is the set of propositional variables; in the recursive definition of a valuation, the values of all propositional variables must therefore be specified. When the truth values of the propositional variables are given, the truth values of all the other propositional formulas are automatically determined from the recursive step of the definition.

Q **Valuation.** We can define a particular valuation in the following way: Let t be an assignment of truth values to propositional variables such that P and Q receive the value **1**, and all other propositional variables receive the value **0**. Let v be recursively defined as above for all other propositional formulas, so that v becomes a valuation. Now v is a function from propositional variables to $\{0,1\}$ that meets the requirements of being a valuation. For example, we can calculate that $v(P \wedge Q) = \mathbf{1}$ because $v(P) = \mathbf{1}$ and $v(Q) = \mathbf{1}$, and that $v(P \wedge R) = \mathbf{0}$ because $v(R) = \mathbf{0}$. ♦

Lists

We will here look at how we can define functions recursively on the inductively defined set of lists from Chapter 9 (*page 103*).

Q **Length of a list.** We want to define a function, LEN, from lists to natural numbers such that if L is a list, then LEN(L) returns the **length** of L, that is, the number of elements of L. We define LEN recursively in the following way:

(1) LEN$(()) = 0$,
(2) LEN$(x :: L) = 1 + $ LEN(L).

For example, we can determine that LEN$((a, b, c)) = 3$ by the following calculation:

$$
\begin{aligned}
\text{LEN}((a, b, c)) && \\
&= 1 + \text{LEN}((b, c)) && \text{by (2) in the definition of LEN} \\
&= 1 + 1 + \text{LEN}((c)) && \text{by (2) in the definition of LEN} \\
&= 1 + 1 + 1 + \text{LEN}(()) && \text{by (2) in the definition of LEN} \\
&= 1 + 1 + 1 + 0 && \text{by (1) in the definition of LEN} \\
&= 3 && \text{by ordinary calculation} \quad ♦
\end{aligned}
$$

Q **Concatenation of lists.** We will now define a more general way of **appending** or **concatenating** lists, more specifically an operation + such that $(1, 2, 3) + (4, 5, 6) = (1, 2, 3, 4, 5, 6)$. The following is a recursive definition of +:

(1) $() + M = M$,
(2) $(x :: L) + M = x :: (L + M)$.

Notice that the first argument of + becomes *smaller* in (2). The left-hand side has $(x :: L)$, but on the right, we have only L. The second argument is always M. Here is an example of how + works:

$$(1, 2, 3) + (4, 5, 6)$$

$= 1 :: ((2, 3) + (4, 5, 6))$	by (2) in the definition of $+$
$= 1 :: (2 :: ((3) + (4, 5, 6)))$	by (2) in the definition of $+$
$= 1 :: (2 :: (3 :: (() + (4, 5, 6))))$	by (2) in the definition of $+$
$= 1 :: (2 :: (3 :: (4, 5, 6)))$	by (1) in the definition of $+$
$= 1 :: (2 :: (3, 4, 5, 6))$	by the definition of $::$
$= 1 :: (2, 3, 4, 5, 6)$	by the definition of $::$
$= (1, 2, 3, 4, 5, 6)$	by the definition of $::$ ◆

Q   Image of a list. Here we define a function MAP that takes two arguments, a function f and a list L, and that returns a new list, called the **image** of L under f. In the image of L under f, each element x has been replaced with $f(x)$:

(1) $\text{MAP}(f, ()) = ()$,
(2) $\text{MAP}(f, x :: L) = (f(x) :: \text{MAP}(f, L))$.

For example, if f is the function that multiplies a number by four, we get the following calculation:

$$\text{MAP}(f, (1, 2, 3))$$

$= 4 :: \text{MAP}(f, (2, 3))$	by (2) in the definition of MAP
$= 4 :: (8 :: \text{MAP}(f, (3)))$	by (2) in the definition of MAP
$= 4 :: (8 :: (12 :: \text{MAP}(f, ())))$	by (2) in the definition of MAP
$= 4 :: (8 :: (12 :: ()))$	by (1) in the definition of MAP
$= 4 :: (8 :: (12))$	by the definition of $::$
$= 4 :: (8, 12)$	by the definition of $::$
$= (4, 8, 12)$	by the definition of $::$ ◆

Binary Trees

In Chapter 9 (*page 104*) we defined binary trees inductively, and we may now therefore define functions on them recursively.

Q   Number of nodes in a binary tree. Let the function NUMBER from binary trees to the natural numbers be defined in the following way. The function returns the *number of nodes* in the binary tree:

(1) $\text{NUMBER}(()) = 0$,
(2) $\text{NUMBER}((L, x, R)) = \text{NUMBER}(L) + 1 + \text{NUMBER}(R)$.

This gives that the number of nodes in a leaf node $((\,),x,(\,))$ equals 1:

$$\text{NUMBER}(((\,),x,(\,))) = \text{NUMBER}((\,)) + 1 + \text{NUMBER}((\,)) = 0 + 1 + 0 = 1.$$

If we draw the nodes in the tree as dots, we can write this as $\text{NUMBER}(\bullet) = 1$. Now we can calculate the number of nodes in the binary tree ❦ in this way:

$$\text{NUMBER}(\text{❦}) = \text{NUMBER}(\bullet) + 1 + \text{NUMBER}(\bullet) = 1 + 1 + 1 = 3.$$

We see that the answer is correct because there are three nodes in ❦ . We can calculate a more complex binary tree like this:

$$\text{NUMBER}(\text{❦}) = \text{NUMBER}(\text{❦}) + 1 + \text{NUMBER}(\text{❦}) = (1 + 1 + 0) + 1 + 3 = 6. \qquad \blacklozenge$$

Q **The height of a binary tree.** Let the **height** of a binary tree be defined in the following way, as a recursively defined function HEIGHT from binary trees to the natural numbers. The function returns the *number of nodes* in a branch of maximal length of the binary tree:

(1) $\text{HEIGHT}((\,)) = 0$,

(2) $\text{HEIGHT}((V,x,H)) = 1 + \max(\text{HEIGHT}(V), \text{HEIGHT}(H))$.

As in the previous example, we get that $\text{HEIGHT}(\bullet) = 1$, and we can calculate the height of the binary tree ❦ in this way:

$$\text{HEIGHT}(\text{❦}) = 1 + \max(\text{HEIGHT}(\bullet), \text{HEIGHT}(\bullet)) = 1 + \max(1, 1) = 1 + 1 = 2. \qquad \blacklozenge$$

Formal Languages

Q **Recursively defined function on a formal language.** In Chapter 9 (*page 106*) we defined the language $a^n b a^n$ as the smallest set S such that $b \in S$ and if $t \in S$, then $ata \in S$. We can now, for example, define a function $f : S \to \mathbb{N}$ recursively such that $f(b) = 0$ and $f(ata) = 2 + f(t)$. We calculate what f does on the string aaabaaa in this way:

$$\begin{aligned}
f(\text{aaabaaa}) &= 2 + f(\text{aabaa}) \\
&= 2 + 2 + f(\text{aba}) \\
&= 2 + 2 + 2 + f(\text{b}) \\
&= 2 + 2 + 2 + 0 = 6.
\end{aligned}$$

The function returns the number of a's in the argument. $\qquad \blacklozenge$

Q **Length of a symmetric string.** In Chapter 9 (*page 106*), we also defined the set of symmetric strings with an even number of characters over $\{a, b\}$. The following is a recursively defined function that calculates the *length* of such a string:

– Let $L(\Lambda) = 0$,
– if x is a symmetric string, then $L(axa) = L(x) + 2$ and $L(bxb) = L(x) + 2$.

We can calculate what L does with the argument abaaba in the following way:

$$
\begin{aligned}
L(abaaba) &= L(baab) + 2 \\
&= (L(aa) + 2) + 2 \\
&= ((L(\Lambda) + 2) + 2) + 2 \\
&= ((0 + 2) + 2) + 2 \\
&= (2 + 2) + 2 = 4 + 2 = 6 \,.
\end{aligned}
$$

◆

Q **Concatenation of strings.** The set of strings over an alphabet A is the smallest set A^* such that

– $\Lambda \in A^*$, and
– if $s \in A^*$ and $x \in A$, then $sx \in A^*$.

We can define the **concatenation** of strings recursively in the following way. To make it clear, we now write $(t \cdot s)$ for the concatenation of t and s, but normally we just write ts. We define the function recursively without doing anything with t, and we can therefore assume that t is an arbitrary string:

– $(t \cdot \Lambda) = t$, and
– if $s \in A^*$ and $x \in A$, then $(t \cdot sx) = (t \cdot s)x$.

Note that the recursive definition has the same structure as the inductive definition. We get that $(ab \cdot ab) = (ab \cdot a)b = (ab \cdot \Lambda)ab = abab$.

◆

Recursion and Programming

In computer science and programming, recursion is a powerful and potentially effective tool. In most programming languages it is possible to define functions and methods recursively, which means that we can formulate algorithms and procedures that solve a problem by dividing it into smaller problems. For example, a program may split a list in two, continue in the same way with each half, and then combine the results at the end. This is a form of the "divide and conquer" technique. When we program in this way, we define functions that "call themselves," but this is completely unproblematic as long as the functions eventually end up in a *base set*. In the programming language *Scheme*, we can, for example, define the factorial function in the following way:

```
(define (factorial x)
   (if (= x 0)
      1
      (* x (factorial (- x 1)))))
```

Exercises

10.1 Here are some different ways to define a function recursively. In all cases we have
$f(0) = 1$. Find the values of $f(1)$, $f(2)$, $f(3)$, and $f(4)$.

(a) $f(n + 1) = f(n) + 3$ (d) $f(n + 1) = 2^{f(n)}$

(b) $f(n + 1) = 3f(n)$ (e) $f(n + 1) = f(n)^2 + f(n) + 1$

(c) $f(n + 1) = -2f(n)$ (f) $f(n + 1) = -f(n)$

10.2 Define the following functions on $\mathbb{N}$ recursively. The function $f(n) = 2n$ may, for
example, be defined recursively by letting $f(0) = 0$ and $f(n+1) = f(n) + 2$. Calculate
the first eight values for each function.

(a) $f(n) = 7n$ (d) $f(n) = 1089$ (g) $f(n) = n(n + 1)$

(b) $f(n) = 2n + 1$ (e) $f(n) = 4n + 2$ (h) $f(n) = n^2$

(c) $f(n) = 10^n$ (f) $f(n) = 1 + (-1)^n$ (i) $f(n) = n^3$

10.3 (a) Define a function LEN recursively on bit strings that provides the *length* of a bit
string, that is, the number of characters in it. For example, LEN$(0101) = 4$.

(b) Define a function REMOVE recursively on bit strings that removes all the 0's in
the string. For example REMOVE$(01001010) = 111$.

(c) How can these functions be combined into a function that gives the number of
1's in a string?

(d) Define a function recursively on bit strings that gives the number of 1's in the
string.

10.4 What is wrong with defining a function "recursively" by saying that $f(0) = 0$ and
$f(x) = f(x) + 2$?

10.5 We have seen how the Fibonacci numbers are defined recursively by saying that
each number $F(n + 2)$ equals the sum of the two previous numbers. Let us do
the same for bit strings, and let each new string be equal the concatenation of the
previous two strings. Let the function $\mathbf{F}$ from the natural numbers to bit strings be
defined recursively by the following conditions:

$$\mathbf{F}(0) = 0 \quad \text{and} \quad \mathbf{F}(1) = 01 \quad \text{and} \quad \mathbf{F}(n) = \mathbf{F}(n - 1)\mathbf{F}(n - 2).$$

(a) Calculate $\mathbf{F}(n)$ for all n up to 6.

(b) What are the lengths of the strings? Can you define what the length of $\mathbf{F}(n)$ is?

(c) How many 0's and 1's are there in the strings? What is the connection between
this and the Fibonacci numbers?

10.6 When a function f is defined recursively, the name of f is used in the definition.
Explain why this is *not* a problem. Must not f already be defined in order to do this?

10.7   Define a function s from propositional formulas to the natural numbers recursively such that if F is a propositional formula, then $s(F)$ equals the *number of symbols* in F. Hint: You must be careful to include all the parentheses. Here are some examples: $s(P) = 1$, $s(\neg P) = 2$, $s((P \rightarrow Q)) = 5$, and $s(((P \rightarrow Q) \wedge \neg R)) = 10$.

10.8   Let A and B be alphabets. Define a function PAR that takes two arguments, a list over A and a list over B of equal length, and returns a list over $A \times B$ as its value. The function returns a list of tuples, where each tuple is a pair of elements from the lists. Hint: If $(x :: M)$ and $(y :: N)$ are the arguments of the function, then the first pair in the new list is $\langle x, y \rangle$. Here are some examples:

$$\text{PAR}((), ()) = ()$$
$$\text{PAR}((a), (1)) = (\langle a, 1 \rangle)$$
$$\text{PAR}((b), (2)) = (\langle b, 2 \rangle)$$

$$\text{PAR}((a, b), (1, 2)) = (\langle a, 1 \rangle, \langle b, 2 \rangle)$$
$$\text{PAR}((1, 2), (3, 4)) = (\langle 1, 3 \rangle, \langle 2, 4 \rangle)$$
$$\text{PAR}((1, 1), (2, 2)) = (\langle 1, 2 \rangle, \langle 1, 2 \rangle)$$

(a) Define the function PAR recursively.

(b) Define the functions FIRST and SECOND recursively that do the opposite of PAR by "unpacking" the tuples in the following way:

$$\text{FIRST}((\langle a, 1 \rangle)) = (a)$$
$$\text{FIRST}((\langle 1, 3 \rangle, \langle 2, 4 \rangle)) = (1, 2)$$

$$\text{SECOND}((\langle a, 1 \rangle)) = (1)$$
$$\text{SECOND}((\langle 1, 3 \rangle, \langle 2, 4 \rangle)) = (3, 4)$$

(c) Is it always true that $\text{FIRST}(\text{PAR}(M, N)) = M$ and $\text{SECOND}(\text{PAR}(M, N)) = N$?

10.9   Let A be an alphabet. Assume that the function ISIN takes two arguments, a character from A and a list over A, and returns either **1** or **0**. Suppose that ISIN behaves in the following way:

$$\text{ISIN}(1, (1, 2)) = 1 \qquad \text{ISIN}(2, (1, 2, 3)) = 1 \qquad \text{ISIN}(1, ()) = 0$$
$$\text{ISIN}(3, (1, 2)) = 0 \qquad \text{ISIN}(4, (1, 2, 3)) = 0 \qquad \text{ISIN}(1, (1)) = 1$$

(a) What does the function do?   (b) Define the function recursively.

10.10   Define a function on binary trees that counts the number of leaf nodes.

10.11   Here are three recursively defined functions on binary trees that "flatten" a tree and return a list in three different ways:

$$\text{PREFIX}(()) = (), \quad \text{PREFIX}((V, x, H)) = (x) + \text{PREFIX}(V) + \text{PREFIX}(H),$$
$$\text{INFIX}(()) = (), \quad \text{INFIX}((V, x, H)) = \text{INFIX}(V) + (x) + \text{INFIX}(H),$$
$$\text{POSTFIX}(()) = (), \quad \text{POSTFIX}((V, x, H)) = \text{POSTFIX}(V) + \text{POSTFIX}(H) + (x).$$

Let $T_1 = ((((), 3, ()), 2, ((), 4, ())), 1, (((), 6, ()), 5, ((), 7, ())))$. We can draw T_1 as in the figure to the right.

```
3   4     6   7
 \ /       \ /
  2         5
    \      /
       1
```

(a) Calculate $\text{PREFIX}(T_1)$, $\text{INFIX}(T_1)$, and $\text{POSTFIX}(T_1)$.

(b) Explain in your own words what the functions do.

(c) Suppose that $\text{PREFIX}(T_2) = (1, 2, 3, 4)$. What can T_2 be?

(d) Suppose that $\text{POSTFIX}(T_3) = (1, 2, 3, 4)$. What can T_3 be?

Chapter 11

Mathematical Induction

■■■■■■■■■■■■■╻■■■■■■■■■■■■

In this chapter you will learn about mathematical induction, a powerful and useful method of proof for statements about natural numbers. You will see many examples of proof by induction and learn about how recursively defined functions and proofs by induction are related. You will also learn about the mathematical game Tower of Hanoi and how it can be analyzed using recursively defined functions and proofs by induction.

A Mathematical Experiment

Let us add consecutive odd numbers and see what we get.

$$1 = 1$$
$$1 + 3 = 4$$
$$1 + 3 + 5 = 9$$
$$1 + 3 + 5 + 7 = 16$$
$$1 + 3 + 5 + 7 + 9 = 25$$
$$1 + 3 + 5 + 7 + 9 + 11 = 36$$

It seems that the sums are *perfect squares*, that is, numbers of the form $n \cdot n$. On closer inspection, it also looks like the sum of the first n odd numbers is equal to n^2:

$$\underbrace{1 + 3 + 5 + \cdots + (2n - 1)}_{\text{the } n \text{ first odd numbers}} = n^2 \,.$$

The question is now, how we can *prove* that this is true for all natural numbers?

Mathematical Induction

One answer is *mathematical induction*, a proof method for showing that something is true for all natural numbers. This proof method is used extensively in mathematics and computer science. In the next chapter, we will generalize this method and look at *structural induction*, a proof method used for showing that something is true for all elements of an inductively defined set. We can look at mathematical induction as a special case of structural induction. Induction is generally a powerful and useful method to show that all the elements of an infinite set have a particular property.

© Springer Nature Switzerland AG 2021
R. Antonsen, *Logical Methods*, https://doi.org/10.1007/978-3-030-63777-4_12

Definition 11.1. Mathematical induction

To prove that a statement is true for all natural numbers, it is sufficient to prove the following:

- The **base case**: that the statement holds for the number 0.
- The **inductive step**: that *if* the statement holds for an arbitrary natural number n, *then* it also holds for $n + 1$. The assumption that the statement is true for n is called the **induction hypothesis**.

If both of these hold, we can conclude by **mathematical induction** that the statement is true for *all* natural numbers.

Q   Dominoes. Suppose we have an infinite row of dominoes, labeled $0, 1, 2, 3, \ldots$, that are standing upright next to each other:

Let $D(n)$ mean that "domino n falls," and suppose the first domino falls, that is, that $D(0)$ is true. This is the *base case*, and we can illustrate it in the following way:

Any domino that falls will make the next one fall, that is, if $D(n)$ is true, then $D(n + 1)$ is true. This is the *inductive step*, and we can illustrate it as follows:

By mathematical induction it follows that all dominoes fall. We have that $D(0)$ is true and that $D(n) \rightarrow D(n+1)$ is true for all natural numbers n, and so it follows by induction that $D(n)$ is true for all natural numbers n. We can illustrate this as follows:

Notation. If P stands for a statement involving a symbol x, it is common to write it as $P(x)$. That makes it easy to substitute something else for x. The symbol x here functions as a *placeholder*. If, for example, $P(x)$ is the statement "x plus x equals $2 \cdot x$," then $P(5)$ is the statement "5 plus 5 equals $2 \cdot 5$."

With the notation above, we can write the two steps of a proof by induction of a statement P – the base case and the inductive step – in the following way:

- The *base case*: $P(0)$ is true.
- The *inductive step*: $P(n) \rightarrow P(n+1)$ is true for all natural numbers n.

If these two assertions are satisfied, the statement must be true for all natural numbers. *Convince yourself that this is true. How could these assertions be fulfilled without the statement being true for all natural numbers?*

To show that $P(n) \rightarrow P(n+1)$ is true for all n is the same as showing that *if* $P(n)$ is true for an arbitrary n, *then* $P(n+1)$ is also true. Note that we do not assume that $P(n)$ is true for all n. That is what we conclude at the end. What we are aiming to show is that $P(n+1)$ is true under the *assumption*, or *hypothesis*, that $P(n)$ is true. That is why it is called the *induction hypothesis*.

Also note how mathematical induction exactly follows the structure of how the natural numbers are inductively defined. And it is always *universal statements* that are proved by induction; in this case, that the statement holds *for all* natural numbers n. The statement in question may be composite, and it is often beneficial to go back to check exactly what it is says.

Note. If the statement we prove by induction does not make sense for, or does not apply to, the number 0, and we want to talk about *positive integers* instead of natural numbers, the base case may be changed so that we begin with the number 1 instead of 0. Then the statement will hold only for all positive integers, but the method works in the same way.

Back to the Experiment

The statement $P(n)$ we wish to prove is

$$\underbrace{1 + 3 + 5 + \cdots + (2n - 1)}_{\text{the first } n \text{ odd numbers}} = n^2 .$$

It is natural to start with the number 1 in this case. To establish the *base case*, that $P(1)$ is true, we replace n with 1 and get the following statement:

$$\underbrace{(2 \cdot 1 - 1)}_{\text{the first odd number}} = 1^2 .$$

By calculating the left and right sides, we get 1 on each side, and so we see that the statement is true for the number 1. Now we will prove the *inductive step*, that if $P(n)$ is true, then $P(n+1)$ is true, for all positive integers n. Therefore, assume that $P(n)$ is true for an arbitrary n:

$$\underbrace{1 + 3 + 5 + \cdots + (2n - 1)}_{\text{the first } n \text{ odd numbers}} = n^2 .$$

This is the statement $P(n)$, and this is our *induction hypothesis*. It is from *this assumption* that we must show that $P(n+1)$ is true:

$$\underbrace{1+3+5+\cdots+(2n-1)+(2n+1)}_{\text{the first } (n+1) \text{ odd numbers}} = (n+1)^2.$$

This is the statement $P(n+1)$, that is, the *statement* for the number $n+1$. Because we have assumed that the statement is true for the number n, we can replace the first part, $1+3+5+\cdots+(2n-1)$, with n^2. Then we get the following:

$$n^2 + (2n+1) = (n+1)^2.$$

If we multiply out the right-hand side, we get that the left- and right-hand sides are identical:

$$n^2 + 2n + 1 = n^2 + 2n + 1.$$

We can conclude that the statement is true for $n+1$. We have thus proved that if $P(n)$ is true, then $P(n+1)$ is true.

By mathematical induction it follows that the statement is true for all positive integers. Now we have proved, for all time and with 100 percent certainty, that the sum of the first n odd numbers is a perfect square, more precisely, n^2.

A Geometric Proof of the Same Claim

We have proved that the sum of successive odd numbers beginning with 1 is a perfect square using mathematical induction, but it is also possible to prove this in many other ways. The following picture is a *geometric* argument for the same statement. Each time an odd number is added, we get another square. See whether you can see the connection between this image and the proof by induction.

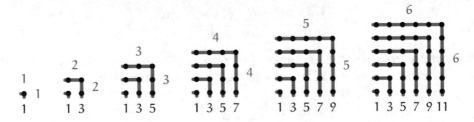

What Really Goes On in an Induction Proof?

When we prove the inductive step, that "if $P(n)$ is true, then $P(n+1)$ is true" for all natural numbers n, we essentially get infinitely many mini proofs, one for each natural number n. For example, if we replace n with 5, we get a proof of $P(5) \rightarrow P(6)$, and if we replace n with 6, we get a proof of $P(6) \rightarrow P(7)$. If we want to prove that the statement P holds for the number 27, can we compose the proofs for $P(0)$ with all the mini proofs for $n = 0, 1, 2, \ldots, 26$:

- The proof of P(0) and the proof of P(0) → P(1) gives a proof of P(1).
- The proof of P(1) and the proof of P(1) → P(2) gives a proof of P(2).
- The proof of P(2) and the proof of P(2) → P(3) gives a proof of P(3).
  ...
- The proof of P(26) and the proof of P(26) → P(27) gives a proof of P(27).

No matter how big the number n is, we can find a proof of P(n). This is exactly what a proof by induction *generalizes*. The conclusion in the proof by induction is that the statement holds for *all natural numbers* n.

Trominoes

A well-known puzzle is whether it is possible to completely cover a chessboard with *dominoes* if we remove two outside squares diagonally opposite each other. We assume that one domino covers exactly two neighboring squares.

The task turns out to be difficult: we are always left with two squares.

❷ Prove that it is impossible. Hint: A chessboard has black and white squares.   ♦

Now we are going to look at a similar question for *trominoes* of the form . This will give us a beautiful example of mathematical induction. We are going to limit ourselves to looking at boards with sizes $2 \times 2, 4 \times 4, 8 \times 8$, etc., and examine whether it is possible to completely cover these with trominoes if we remove one of the squares. First we check the simplest board, with 2×2 squares:

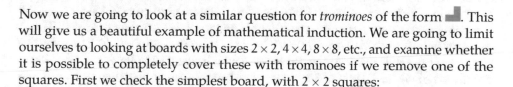

We see that we are able to cover the board no matter which square is removed. We proceed to check the boards with 4×4 squares, and here there are sixteen possibilities for which square to remove. Instead of checking by hand, we can use our imagination. A board with 4×4 squares can be divided into four equal subboards, each with 2×2 squares. When we remove one square, it has to lie in one of these subboards. But then we know – from our analysis of boards with 2×2 squares – that *that* subboard can be covered with trominoes. But what about the other three subboards? In these *no* square is removed! Here comes the trick: Place one tromino in the middle of the large board so that it uses exactly one free square from each of the remaining subboards. There is always a way to do this, and for each of the subboards, this will be just as if one of the squares had been removed. Then we know – again from our analysis of boards with 2×2 squares – that these three parts can be covered with trominoes. Here are two different concrete situations that can arise. In (a), one square is removed; in (b), we solve the 2×2 board with a square removed, in (c), we add an extra piece; and in (d), we solve the other three 2×2-boards:

(a)   (b)   (c)   (d)   (a)   (b)   (c)   (d)

We conclude that *all* boards that have 4×4 squares may be completely covered, no matter which square is removed. The great thing about this strategy is that it can be *generalized*. A board with 8×8 squares can also be divided into four equally large subboards such that each subboard has 4×4 squares. If we remove one square, it must be in one of these four subboards, which can now be solved as above. We place a tromino in the middle of the big board such that it uses exactly one square from each of the remaining subboards. We can continue in this way:

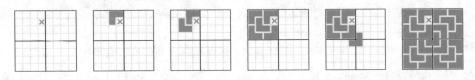

We prove the statement "each board with $2^n \times 2^n$ squares with one square removed can be completely covered with trominoes" for all n by mathematical induction: the *base case* in the proof is that the statement holds for $n = 1$. There are four different cases for boards with 2×2 squares, and these we have already checked. The *inductive step* is to show that if the statement holds for n, it also holds for $n + 1$. But this is exactly what we have shown: Suppose that a board with $2^n \times 2^n$ squares with one square removed can be completely covered with trominoes. This is our *induction hypothesis*. If we take a board with $2^{n+1} \times 2^{n+1}$ squares, we can use the same trick. *By mathematical induction it follows that the statement is true for all boards with $2^n \times 2^n$ squares.*

Properties of Recursively Defined Functions

Mathematical induction can be used to show that recursively defined functions have specific properties. The following example illustrates this.

Q Quadrupling. Let $f : \mathbb{N} \to \mathbb{N}$ be the function recursively defined by

  − $f(0) = 0$,
  − $f(n + 1) = f(n) + 4$, for all $n \in \mathbb{N}$.

It looks like the function f quadruples its argument. Let us try to show by induction that $f(n) = 4n$ for all $n \in \mathbb{N}$. Let $P(n)$ stand for $f(n) = 4n$.

The *base case* consists in showing that $P(0)$ is true, meaning that $f(0) = 4 \cdot 0$ is true. By the definition of f, we have $f(0) = 0$, and $4 \cdot 0 = 0$, so $P(0)$ is true.

The *inductive step* consists in showing that *if* $P(n)$ is true, *then* $P(n + 1)$ is true. Therefore, suppose that $P(n)$ is true for an arbitrary number n, that is, that $f(n) = 4n$. This is the *induction hypothesis*. The goal now is to show that $P(n + 1)$ is true, that is, that $f(n + 1) = 4(n + 1)$. But this we can calculate. The definition of f gives that $f(n + 1) = f(n) + 4$. Because of the *induction hypothesis*, which says that $f(n) = 4n$, this equals $4n + 4$. If we factor out the 4, we get $4(n + 1)$, and we are done. We can set out this argument in the following way:

$$f(n + 1) = f(n) + 4 \qquad \text{by the definition of f}$$
$$= 4n + 4 \qquad \text{by the induction hypothesis}$$
$$= 4(n + 1) \qquad \text{by ordinary calculation}$$

We have thus calculated that $f(n + 1) = 4(n + 1)$, which means that $P(n + 1)$ is true. *By induction, it follows that the statement is true for all natural numbers.* ◆

Digression

Here is a "proof by induction" that all juggling balls have the same color. *See whether you can find the error!* The proof is by induction on the set of all juggling balls. Let the statement we wish to prove for all n be "in every set of n juggling balls, all balls have the same color." The *base case* is that the statement holds for 1, and that is correct, because in all sets with one juggling ball, all balls have the same color. The *inductive step* consists in showing that if the statement holds for n, it holds for $n + 1$. Suppose the statement holds for n, meaning that "in every set of n juggling balls, all balls have the same color." This is the *induction hypothesis*. We must show that *on that assumption*, the statement holds for $n + 1$, that is, that "in every set of $n + 1$ juggling balls, all balls have the same color."

n juggling balls

n+1 juggling balls

Now let B be an arbitrary set with $n + 1$ juggling balls. If we look at a subset of B with n balls, it follows from the induction hypothesis that all of these have the same color, and if we look at another subset of B with n balls, it follows from the induction hypothesis that these too have the same color. If we place all the balls in a row, we can look at the n first and the n last balls:

n juggling balls

n juggling balls

The sets have to have something in common, and then it follows that all the $n + 1$ balls in B must have the same color. *By induction it follows that the statement holds for all n.* That is, all juggling balls have the same color.

The Tower of Hanoi

The mathematical game **Tower of Hanoi** was popularized in 1883 by the French mathematician *Édouard Lucas* (1842–1891) and is related to many basic problems in mathematics and computer science. The game consists of disks of different sizes, and it begins with at all the disks placed on one of the rods and arranged such that every disk is resting on a larger disk:

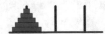

The goal is to move all the disks over to another rod. We are permitted to move exactly one disk at a time, one of those at the top, and it is not allowed to place a disk on top of a smaller disk. The question is now, *What is the least number of moves needed to move all the disks from one rod to another?* It is a good idea to look at the smallest cases first. The *smallest* case is zero disks, and then we need zero moves. If we have one disk, the answer is *one move*. If we have two disks, we can solve the game in the following way:

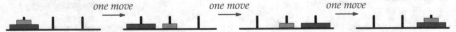

Here we first move the small disk from rod 1 to rod 2, then the big disk from rod 1 to rod 3, and finally the small disk from rod 2 to rod 3. We use *three moves* in total.

Now we let $T(n)$ stand for *the least number of moves needed to move n disks from one rod to another.* We have just shown that $T(1) = 1$ and $T(2) = 3$. Let us look at what happens with three disks:

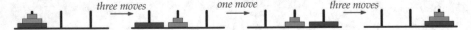

Here we first move two disks from rod 1 to rod 2, in exactly the same way as with only two disks, and this takes *three moves*. Then we move the large disk from rod 1 to rod 3 in *one move*. Finally, we move two disks from rod 2 to rod 3, and this also takes *three moves*. In total, we use *seven moves*, and we can write $T(3) = 7$. We can continue in exactly the same way with four disks:

It takes *seven moves* to move three disks from rod 1 to rod 2, *one move* to move the large disk, and finally *seven moves* to move the three disks in place. That ends up being a total of fifteen moves, and we can write $T(4) = 15$.

There is a beautiful pattern here: In order to move $n + 1$ disks, we need *no more* than $T(n)$ moves to move the n smallest disks, *one move* to move the largest disk, and finally $T(n)$ moves to move the n smallest disks in place. We can express this with the following formula:

$$T(n + 1) \leqslant 2 \cdot T(n) + 1.$$

The reason it says $\leqslant$ here is that we actually have *not* shown that $2 \cdot T(n) + 1$ is the *least* number needed to move $n + 1$ disks. We defined $T(n + 1)$ as *the least number of moves needed to move $n + 1$ disks from one rod to another,* but we have only shown that it is *possible* to move $n + 1$ disks with $2 \cdot T(n) + 1$ moves. It might be that there was an even better way to move the disks!

It turns out that our method actually is the most effective method, but that requires an additional argument: *To move the largest disk, all the other disks must be on the same rod. If not, it is impossible to move it.*

That means that all the other n disks must first be moved away from the largest disk, and that takes *at least* $T(n)$ moves. Afterward, it takes *at least one* move to move the largest disk in place (depending on how many times we move it back and forth …), and then we have to use *at least* $T(n)$ moves to get the other n disks in place. Here we can use several moves, but we must use *at least* $T(n)$ moves. We can express this with the following formula:

$$T(n + 1) \geqslant 2 \cdot T(n) + 1.$$

Now we have shown that we must use *at least* $2 \cdot T(n) + 1$ moves in order to move $n + 1$ disks. But then we have shown that $2 \cdot T(n) + 1$ moves is both *necessary* and *sufficient*, and we can finally write the formula with equality:

$$T(n + 1) = 2 \cdot T(n) + 1.$$

We see that this is a *recursive* function, because the value of $T(n + 1)$ is uniquely determined by the value of $T(n)$. When we also know that $T(0) = 0$, we can figure out values for the other numbers:

n	0	1	2	3	4	5	6	7	8	9	10
$T(n)$	0	1	3	7	15	31	63	127	255	511	1023

It seems that each of these numbers is exactly one less than a power of two. For example, we have that $T(10) = 1023$ and $2^{10} = 1024$. We can therefore *guess* that $T(n)$ equals $2^n - 1$. We have already proved that the property $T(n + 1) = 2 \cdot T(n) + 1$ holds for T, but we have not proved that $T(n) = 2^n - 1$. We do that now. We prove by induction that the statement $T(n) = 2^n - 1$ holds for all natural numbers n.

The *base case* is that the statement holds for $n = 0$, and this is true because both $T(0)$ and $2^0 - 1$ equal zero. The *inductive step* is to show that if the statement holds for n, it also holds for $n + 1$. Therefore, suppose that $T(n) = 2^n - 1$. This is the *induction hypothesis*. From this we must show that $T(n + 1) = 2^{n+1} - 1$:

$$
\begin{aligned}
T(n + 1) &= 2 \cdot T(n) + 1 && \text{by definition of T} \\
&= 2 \cdot (2^n - 1) + 1 && \text{by the induction hypothesis} \\
&= (2^{n+1} - 2) + 1 && \text{by multiplication} \\
&= 2^{n+1} - 1 && \text{by addition}
\end{aligned}
$$

We have proved the inductive step, which is that if the statement holds for n, it also holds for $n + 1$.

By mathematical induction, it follows that the statement is true for all natural numbers.

In other words, we have found two different ways to express the same thing: $T(n)$ is defined as the least number of moves needed to move all the disks from one rod to another. The number $2^n - 1$ is an expression for what this number actually is.

It is worth noting that with 64 disks, this number is

$$2^{64} - 1 = 18\,446\,744\,073\,709\,551\,615\,.$$

If we managed to move one disk per second, it would take us about 600 billion years to finish. This shows how quickly the complexity of something may increase, even with simple rules. This becomes even clearer if we add one or more rods: for *four* rods, the problem becomes so complex that it is still an *open problem* to find the best solution. There is much more to say about the Tower of Hanoi, but we must stop here.

More Summing of Numbers

Q   **The sum of the first n numbers.** We can also take the sum of consecutive natural numbers and see what we get.

$$
\begin{aligned}
1 &= \mathbf{1} \\
1 + 2 &= \mathbf{3} \\
1 + 2 + 3 &= \mathbf{6} \\
1 + 2 + 3 + 4 &= \mathbf{10} \\
1 + 2 + 3 + 4 + 5 &= \mathbf{15} \\
1 + 2 + 3 + 4 + 5 + 6 &= \mathbf{21}
\end{aligned}
$$

These are the same triangular numbers we met in Chapter 10 (*page 111*). After some experimentation, we find that

$$1 + 2 + 3 + \cdots + n = \frac{n(n + 1)}{2}\,.$$

Check that this is true for the first few numbers.

We want to be absolutely sure that this is correct for absolutely *all* positive integers. Therefore, we try a proof by induction. The statement to prove is this:

$$1 + 2 + 3 + \cdots + n = \frac{n(n + 1)}{2}\,.$$

The *base case* is to show that the statement holds for $n = 1$. If we replace n with the number 1, we get the following:

$$1 = \frac{1(1+1)}{2}.$$

After some calculation, we see that this is true, since both the left- and right-hand sides equal 1.

To prove the *inductive step*, we assume that the statement is true for $n = k$. This is the *induction hypothesis*, and it looks like this:

$$1 + 2 + 3 + \cdots + k = \frac{k(k+1)}{2}.$$

It is quite common to use a different variable, like k here, as a placeholder in a proof by induction. Now we have to show that the statement is true for $n = (k+1)$, that is, that the following is true:

$$1 + 2 + 3 + \cdots + k + (k+1) = \frac{(k+1)((k+1)+1)}{2},$$

which is the result of replacing n with $(k+1)$. Because we have assumed that the statement is true for $n = k$, we can replace the first part of the left-hand side, $1 + 2 + 3 + \cdots + k$, with $\frac{k(k+1)}{2}$. Then we get the following:

$$\frac{k(k+1)}{2} + (k+1) = \frac{(k+1)((k+1)+1)}{2}.$$

What is left now is to prove that this is true, that is, that the left- and right-hand sides are the same. We can do that by the following calculation:

$$\begin{aligned}
\frac{k(k+1)}{2} + (k+1) &= \frac{k(k+1)}{2} + \frac{2(k+1)}{2} \\
&= \frac{k(k+1) + 2(k+1)}{2} \\
&= \frac{(k+2)(k+1)}{2} \\
&= \frac{(k+1)(k+2)}{2} \\
&= \frac{(k+1)((k+1)+1)}{2}.
\end{aligned}$$

We can conclude that the statement is true for $n = k+1$. We have thereby proved that *if* the statement is true for k, then the statement true for $(k+1)$.

By mathematical induction, it follows that the statement is true for all positive integers. ◆

Digression

There is a story about the German mathematician *Carl Friedrich Gauss* (1777–1855) when he was in elementary school. The class was given the task of summing up all the numbers from 1 to 100. Almost immediately, he said 5050! *How could he know that?*

1	2	3	4	$\cdots$	97	98	99	100
100	99	98	97	$\cdots$	4	3	2	1
101	101	101	101	$\cdots$	101	101	101	101

The explanation is that above the line appear all the numbers from 1 to 100, but twice: once forward and once backward. If we add the pairs of numbers that are above each other, we always get 101. That is, both above and below the line, the sums of all the numbers are $100 \cdot 101$, but that is twice as much as the number we are looking for. The answer is therefore $(100 \cdot 101)/2$.

Reasoning and Strong Induction

I hope you have seen that mathematical induction is both a *useful* and *powerful* proof method. It is worth noting that although we are proving statements *about natural numbers*, the statements might be about other types of mathematical objects. In the example with the trominoes we proved a statement about every *board* of a certain size, but the statement was formulated about numbers so that we could use mathematical induction.

There are two things we have not discussed in this chapter. One is the *reason* why the method holds. One simple answer is that this is one of our basic assumptions, and I think you will agree that it is a reasonable one. It is common to include *axioms* for mathematical induction in order to make this assumption precise. We could also have *proved* that mathematical induction is a valid principle, but then we would have assumed and used another principle that was equally strong. The other thing we have not discussed is so-called **strong** or **complete** induction. That means that in the *inductive step*, we assume the stronger hypothesis that the statement holds for *all* natural numbers less than or equal to n, and not just n, and from this show that the statement holds for $n + 1$. This is *equivalent* to what we have seen, but it can be advantageous to refer to *all* the foregoing statements and not only the previous one.

In the next chapter we will generalize mathematical induction so that we can prove statements about *all* inductively defined sets.

Exercises

11.1 A set with one element has two subsets, a set with two elements has four subsets, and a set with three elements has eight subsets. Look at the subsets of $\{a, b, c\}$:

$$\varnothing, \ \{a\}, \ \{b\}, \ \{c\}, \ \{a, b\}, \ \{a, c\}, \ \{b, c\}, \ \{a, b, c\}$$

(a) Are all these subsets also subsets of $\{a, b, c, d\}$?

(b) For each of these subsets, there are, in a very natural way, exactly two subsets of $\{a, b, c, d\}$. What are they?

(c) Prove by mathematical induction that a set with n elements has 2^n subsets. Hint: Use the same method as in (a) and (b) above.

11.2 The following is a proof by mathematical induction that the statement

$$n^3 - n \text{ is divisible by } 3$$

is true for all natural numbers $n \geqslant 1$. Figure out what should be in the boxes:

First we show the base $\boxed{^1}$: that $\boxed{^2}$ holds for $n = 1$. We insert for n and check:

$$1^3 - 1 = 0 \text{ is } \boxed{^3} \text{ by } 3.$$

We see that the statement is true for $n = 1$. To prove the $\boxed{^4}$ step, we assume that the statement holds for $n = k$. This is our induction $\boxed{^5}$. We must show from this that the statement holds for $n = k + 1$. That the statement holds for $n = k$ is the same as that $k^3 - k$ is divisible by 3. We can therefore assume that there is a natural number a such that $k^3 - k = 3a$. From this assumption we must show that $\boxed{^6}$ is divisible by 3:

$$
\begin{aligned}
(k+1)^3 - (k+1) &= (k^3 + 3k^2 + 3k + 1) - (k+1) && \text{by multiplying out } (k+1)^3 \\
&= (k^3 - k) + 3k^2 + 3k && \text{by adding and subtracting} \\
&= 3a + 3k^2 + 3k && \text{by the } \boxed{^7} \\
&= 3(a + k^2 + k) && \text{by factoring}
\end{aligned}
$$

Then $(k+1)^3 - (k+1)$ must be divisible by 3, and the statement holds for $n = k + 1$. By $\boxed{^8}$ it follows that the statement is true for $\boxed{^9}$.

11.3 Prove by mathematical induction that $n^2 + n$ is an even number for all natural numbers n.

11.4 Why can't we use mathematical induction to prove statements about fractions or real numbers, just as we do for natural numbers?

11.5   It seems that every third Fibonacci number is an even number:

$$0, \ 1, \ 1, \ 2, \ 3, \ 5, \ 8, \ 13, \ 21, \ 34, \ 55, \ 89, \ 144, \ 233, \ 377, \ 610, \ 987$$

Prove this for all Fibonacci numbers by mathematical induction.

11.6   Let R be a transitive relation. Let aR^nb, for $n \geqslant 1$, mean that there is a sequence of tuples

$$\langle a_0, a_1 \rangle, \langle a_1, a_2 \rangle, \ldots, \langle a_{n-1}, a_n \rangle$$

from R such that $a_0 = a$ and $a_n = b$. Prove the statement "if aR^nb, then aRb" for all natural numbers $n \geqslant 1$ by mathematical induction.

11.7   Check whether 2^n is greater than n for all natural numbers. In that case, can you prove it by mathematical induction?

11.8   Prove by mathematical induction that

$$6^n - 1$$

is divisible by 5 for all natural numbers n.

11.9   Provide a proof by mathematical induction that the following statement holds for all natural numbers n:

$$5 + 9 + 13 + \cdots + (4n + 5) = 2n^2 + 7n + 5.$$

11.10   Provide a proof by mathematical induction that the following statement holds for all natural numbers $n \geqslant 2$:

$$\left(1 - \frac{1}{2}\right)\left(1 - \frac{1}{3}\right) \cdots \left(1 - \frac{1}{n}\right) = \frac{1}{n}.$$

11.11   Provide a proof by mathematical induction that the following statement holds for all natural numbers n:

$$1^2 + 2^2 + 3^2 + \cdots + n^2 = \frac{n(n+1)(2n+1)}{6}.$$

11.12   Let the functions E, for Even number, and O, for Odd number, be defined from the natural numbers to $\{0, 1\}$ as follows:

- $E(0) = 1$ - $E(n+1) = O(n)$
- $O(0) = 0$ - $O(n+1) = E(n)$

(a) Prove that $E(2n) = 1$ and that $O(2n) = 0$ for all natural numbers n.
(b) Prove that $O(2n + 1) = 1$ and that $E(2n + 1) = 0$ for all natural numbers n.

Chapter 12

Structural Induction

■■□□□□□□□□□□□□□▯□□□□□□□□□□□■■

In this chapter you will learn about structural induction, a generalization of mathematical induction that works for all inductively defined sets. You will learn to use this method to prove statements about bit strings, propositional formulas, lists, and binary trees.

Structural Induction

In mathematical induction we are using the *inductive* definition of the set of natural numbers. We will now generalize mathematical induction such that the method can be used for *any* inductively defined set, not just the natural numbers. This more general method of proof, called *structural induction*, is a powerful and versatile method. We can use it when we want to show that all the elements of an inductively defined set have a particular property. A proof by structural induction has a *base case* and an *inductive step* that precisely follow the familiar definition of mathematical induction.

Definition 12.1. Structural induction

Suppose that a set is inductively defined. To show that a statement is true for all elements of this set, it is sufficient to prove the following two assertions:

- The statement holds for all elements of the base set. This step is called the **base case**.
- If the set is closed under an operation that generates x from $x_1, x_2, \ldots, x_n$ *and* the statement holds for all of $x_1, x_2, \ldots, x_n$, then the statement also holds for x. This step is called the **inductive step**. The assumption that the statement holds for $x_1, x_2, \ldots, x_n$ is called the **induction hypothesis**.

If both of these assertions hold, we can conclude by **structural induction** that the statement is true for *all* elements of the set. In that case, we say that we have proved the statement by structural induction on the inductively defined set.

© Springer Nature Switzerland AG 2021
R. Antonsen, *Logical Methods*, https://doi.org/10.1007/978-3-030-63777-4_13

You should compare this definition with the definition of mathematical induction in the previous chapter (*page 124*) and see that it is more general. Also, be aware of the language when talking about proof by induction: long expressions like "by structural induction on the set of formulas" are often abbreviated as "by induction on formulas" or "by induction on the structure of formulas," but they all mean the same thing. With the same notation as before, where $P(x)$ stands for the statement to be proved, we can write the two steps in a proof by structural induction in the following way:

- The *base case*: that $P(x)$ is true for all x in the base set.
- The *inductive step*: that *if* $P(x_1), P(x_2), \ldots, P(x_n)$ are true, and x is generated from $x_1, x_2, \ldots, x_n$, *then* $P(x)$ is also true.

We can imagine structural induction in the following way, where IH stands for the *induction hypothesis*:

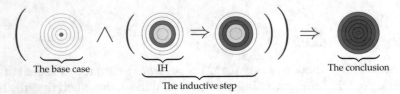

The base case   IH   The conclusion

The inductive step

The *base case* allows us to color what is in the middle. The *induction hypothesis* is that everything in an arbitrarily chosen layer is colored. The *inductive step* uses the induction hypothesis to show that everything in the next layer is also colored. *The conclusion* is that everything is colored.

Structural Induction on Bit Strings

We defined bit strings inductively in Chapter 9 (*page 107*) as the smallest set that contains 0 and 1 such that if b is in the set, then b0 and b1 are also in the set. To show that a statement is true for all bit strings by structural induction, we must prove the base step and the inductive step.

The *base case* consists in showing that the statement is true for all the elements of the base set, namely 0 and 1.

The *inductive step* consists in proving the following two cases, one for each of the operations under which the set is closed:

- For all bit strings b, if the statement is true for b, then it is true for b0.
- For all bit strings b, if the statement is true for b, then it is true for b1.

Note that if we had defined bit strings such that the base set consisted of only the empty string and not 0 and 1 as we have done here, we would have had only one case.

Q   **Structural induction on bit strings.** Let the function v from bit strings to the natural numbers be defined recursively in the following way:

- $v(0) = 0$ and $v(1) = 1$,
- $v(b0) = 2 \cdot v(b)$,
- $v(b1) = 2 \cdot v(b) + 1$.

We can calculate that $v(11) = 2 \cdot v(1) + 1 = 3$ and that the function returns the same value no matter how many instances of 0 we add on the left-hand side: $v(011) = v(0011) = v(00011) = \cdots = 3$. It seems that $v(b) = v(0b)$ holds for all bit strings b. This is what we are now trying to prove by structural induction. We let P(b) stand for the statement $v(b) = v(0b)$.

The *base case* consists in showing that P(0) and P(1) are true:

- P(0) is true: $v(0) = 0 = 2 \cdot 0 = 2 \cdot v(0) = v(00)$,
- P(1) is true: $v(1) = 1 = 2 \cdot 0 + 1 = 2 \cdot v(0) + 1 = v(01)$.

These equalities hold as a result of the definition of v and ordinary rules of calculation.

The *inductive step* consists in showing that if P(b) is true, then so are P(b0) and P(b1). Therefore, suppose that P(b) is true, that is, that $v(b) = v(0b)$, for an arbitrary bit string b; this is the *induction hypothesis*.

- P(b0) is true: $v(b0) = 2 \cdot v(b) = 2 \cdot v(0b) = v(0b0)$,
- P(b1) is true: $v(b1) = 2 \cdot v(b) + 1 = 2 \cdot v(0b) + 1 = v(0b1)$.

For both lines above, the first and last equalities hold by the definition of v, and the middle equality is due to the induction hypothesis, which says that $v(b) = v(0b)$.

By structural induction on the set of bit strings, it follows that $v(b) = v(0b)$ is true for all bit strings b.   ◆

Structural Induction on Propositional Formulas

The set of propositional formulas is defined inductively. Thus we can use *structural induction* to prove that a statement holds for *all* propositional formulas. To show that a statement is true for all propositional formulas, we must prove:

- The *base case*, that the statement is true for all *propositional variables*.
- The *inductive step*, which here consists of four, equally important, parts:
    - If the statement is true for the formula F, it is true for ¬F.
    - If the statement is true for the formulas F and G, it is true for $(F \wedge G)$.
    - If the statement is true for the formulas F and G, it is true for $(F \vee G)$.
    - If the statement is true for the formulas F and G, it is true for $(F \to G)$.

How can we *prove* that an expression $((Q \wedge P)$ is *not* a propositional formula? It is usually much harder to prove that something is *impossible* than that it is possible. It can sometimes be both better, and easier, to prove a *stronger* statement than the one we really want to prove. For example, in this case, we can prove that if an expression has different numbers of left and right parentheses, then it is *not* a propositional formula. The contrapositive of this is that if something is a propositional formula, then it has the same number of left and right parentheses.

This is a stronger statement, and it has as a consequence that $((Q \wedge P)$ is *not* a propositional formula, because it has two left parentheses and one right parenthesis. We prove the stronger statement by structural induction.

Q **Structural induction on propositional formulas.** By structural induction on the set of propositional formulas we prove that all propositional formulas F have the same number of left and right parentheses. The statement we are proving that holds for all formulas F is the following: "F has the same number of left and right parentheses."

The *base case*: If F is a propositional variable, it does not contain parentheses, and therefore the statement is true.

The *inductive step*: Suppose that $F = \neg G$ and that the statement holds for G. The formula F has the same number of left and right parentheses as G. Thus the statement also holds for F. Let F be $(G \circ H)$ for $\circ \in \{\wedge, \vee, \rightarrow \}$, and assume that the statement holds for both G and H. The formula F therefore has one left and one right parenthesis in addition to those in G and H. Because the statement holds for G and H, it also holds for F.

By structural induction on the set of propositional formulas it follows that all propositional formulas F have the same number of left and right parentheses.

A direct consequence of this result is that $((Q \wedge P)$ is *not* a propositional formula, because it does *not* have the same number of left and right parentheses.   ◆

Let us recursively define a function on propositional formulas that replaces a propositional variable with a formula.

Definition 12.2. Substitution

Let P be a propositional variable and H a propositional formula. We recursively define a function that replaces all occurrences of P with H, and we write F[P/H] for the result of **replacing** or **substituting** all P's in F with H:

- $F[P/H] = H$ when F is a propositional variable and $F = P$.
- $F[P/H] = F$ when F is a propositional variable and $F \neq P$.
- $(\neg F)[P/H] = \neg(F[P/H])$.
- $(F \circ G)[P/H] = (F[P/H] \circ G[P/H])$ when $\circ \in \{\wedge, \vee, \rightarrow \}$.

Q **Substitution.** Here are some examples of how the substitution function works:

- $P[Q/(R \rightarrow S)] = P$,
- $Q[Q/(R \rightarrow S)] = (R \rightarrow S)$,
- $(P \rightarrow Q)[Q/(R \rightarrow S)] = (P \rightarrow (R \rightarrow S))$,
- $(Q \vee \neg Q)[Q/(R \rightarrow S)] = ((R \rightarrow S) \vee \neg(R \rightarrow S))$.   ◆

Q **Structural induction on propositional formulas.** We will now prove a fundamental result in logic, which is that we can replace a formula that is contained in a larger formula with an equivalent formula without changing the larger formula's truth value:

Suppose that A and B are equivalent. By structural induction on the set of propositional formulas we shall prove the statement "$F[P/A]$ and $F[P/B]$ are equivalent" for all formulas F.

The *base case* consists in showing that the statement is true for all propositional variables:

- If F is a propositional variable and $F = P$, then $F[P/A]$ equals A and $F[P/B]$ equals B, and they are then equivalent by assumption.
- If F is a propositional variable and $F \neq P$, then $F[P/A]$ and $F[P/B]$ both equal F and are thus equivalent because they are identical.

The *inductive step* consists in showing that if the statement holds for the formulas F and G, it also holds for the composite formulas. Therefore, assume that the statement holds for the formulas F and G. That means that $F[P/A] \Leftrightarrow F[P/B]$ and $G[P/A] \Leftrightarrow G[P/B]$. This is the induction hypothesis. We get four different cases, one for each of the connectives $\neg, \wedge, \vee$, and $\rightarrow$, and we do the case for $\wedge$ here. We must show from the assumption on F and G that $(F \wedge G)[P/A]$ and $(F \wedge G)[P/B]$ are equivalent. Therefore, choose an arbitrary valuation v. Using the following equivalences we show that $(F \wedge G)[P/A]$ and $(F \wedge G)[P/B]$ must have the same value:

$$v((F \wedge G)[P/A]) = 1$$
$$\Leftrightarrow v(F[P/A] \wedge G[P/A]) = 1 \qquad \text{by the definition of substitution}$$
$$\Leftrightarrow v(F[P/A]) = 1 \text{ and } v(G[P/A]) = 1 \qquad \text{by the definition of valuation}$$
$$\Leftrightarrow v(F[P/B]) = 1 \text{ and } v(G[P/B]) = 1 \qquad \text{by the induction hypothesis}$$
$$\Leftrightarrow v(F[P/B] \wedge G[P/B]) = 1 \qquad \text{by the definition of valuation}$$
$$\Leftrightarrow v((F \wedge G)[P/B]) = 1 \qquad \text{by the definition of substitution}$$

This means that the formulas are equivalent. The proofs for the other connectives are done in a similar way.

By structural induction it follows that the statement holds for all propositional formulas. ♦

Structural Induction on Lists

We already have defined lists in Chapter 9 (*page 103*) and the functions LEN, +, and MAP on lists in Chapter 10 (*pages 117 and 118*),and we will now see how we can use structural induction to prove statements about lists and functions on them. We do it in the form of some exercises. Try to solve them yourself before reading on.

❷ Prove by induction on lists that $(L + ()) = (() + L)$ for all lists L.   ♦

❶ The *base case* is that the statement holds if L is replaced with the empty list. This is true because $(() + ()) = ()$.

The *inductive step*: Suppose that the statement holds for the list L, that is, that $(L + ()) = (() + L)$. This is the induction hypothesis. From this we must show that the statement holds for $(x :: L)$, that is, that $(x :: L) + () = () + (x :: L)$. We can do this as follows:

$$
\begin{aligned}
(x :: L) + () &= x :: (L + ()) && \text{by part (2) of the definition of +} \\
&= x :: (() + L) && \text{by the induction hypothesis} \\
&= x :: L && \text{by part (1) of the definition of +} \\
&= () + (x :: L) && \text{by part (1) of the definition of +}
\end{aligned}
$$

By structural induction, it follows that $L + () = () + L$ *for all lists L.* ◆

❷ Prove by induction on lists that $\text{LEN}(M + N) = \text{LEN}(M) + \text{LEN}(N)$ for all lists M and N. Hint: Let N be a placeholder for an arbitrary list and look at the structure of M.◆

❸ We show that the statement holds no matter what N stands for and let N stand for an arbitrary list. In the following proof, we are looking only at the structure of M and not on how N is constructed. This is sometimes called "doing induction on the structure of M."

The *base case* is that the statement holds if M is replaced with the empty list. We get the following calculation, which shows that $\text{LEN}(() + N) = \text{LEN}(()) + \text{LEN}(N)$:

$$
\begin{aligned}
\text{LEN}(() + N) &= \text{LEN}(N) && \text{by part (1) of the definition of +} \\
&= 0 + \text{LEN}(N) && \text{by ordinary calculation} \\
&= \text{LEN}(()) + \text{LEN}(N) && \text{by part (1) of the definition of LEN}
\end{aligned}
$$

The *inductive step*: Suppose that the statement holds for M, that is, that $\text{LEN}(M+N) = \text{LEN}(M) + \text{LEN}(N)$. This is the induction hypothesis. From this we must show that the statement holds for $(x :: M)$, that is, that $\text{LEN}((x :: M) + N) = \text{LEN}(x :: M) + \text{LEN}(N)$. We can do this in the following way:

$$
\begin{aligned}
\text{LEN}((x :: M) + N) &\\
&= \text{LEN}(x :: (M + N)) && \text{by part (2) of the definition of +} \\
&= 1 + \text{LEN}(M + N) && \text{by part (2) of the definition of LEN} \\
&= 1 + \text{LEN}(M) + \text{LEN}(N) && \text{by the induction hypothesis} \\
&= \text{LEN}(x :: M) + \text{LEN}(N) && \text{by part (2) of the definition of LEN}
\end{aligned}
$$

By structural induction, it follows that $\text{LEN}(M + N) = \text{LEN}(M) + \text{LEN}(N)$ *for all lists M and N.* ◆

Structural Induction on Binary Trees

We defined binary trees in Chapter 9 (*page 104*) and the functions NUMBER and HEIGHT on binary trees in Chapter 10 (*pages 118 and 119*). We will now see how we can use structural induction to prove some simple statements about these.

Q Structural induction on binary trees. As before, we let • represent a node in a binary tree. Using the definitions of NUMBER and HEIGHT, we can calculate the following values:

T	( )	•	⋰	⋎	⋱	⦃	⋏	⦄	⋎	⋎	⋎⋎
NUMBER(T)	0	1	2	3	3	3	4	4	5	5	7
HEIGHT(T)	0	1	2	2	3	3	3	4	3	3	3

Looking at the values and the binary trees in the table, we see that the function NUMBER returns the *number of nodes* in each of the binary trees. This is a simple property of the function NUMBER, and we will – for the sake of training – prove by structural induction that this always holds. The statement we prove for all binary trees T is thus "NUMBER(T) equals the number of nodes in T."

The *base case* consists in showing that the statement holds for the empty binary tree ( ). The number of nodes in the empty binary tree is zero, and the statement holds because NUMBER(( )) = 0 by the definition of NUMBER.

The *inductive step* consists in showing that *if* the statement holds for two binary trees L and R, *then* it also holds for the tree (L, x, R). Therefore, assume that the statement holds for the trees L and R, that is, that the number of nodes in L equals NUMBER(L) and that the number of nodes in R equals NUMBER(R). The number of nodes in the tree (L, x, R) equals the sum of the numbers of nodes in L and R plus one. But that is exactly how NUMBER is defined:

$$\text{NUMBER}((L, x, R)) = \text{NUMBER}(L) + 1 + \text{NUMBER}(R).$$

Thus the number of nodes in (L, x, R) equals NUMBER((L, x, R)).

By structural induction it follows that the function NUMBER returns the number of nodes in the tree.

Some comments are in place. You may think this was unnecessarily complicated, but it is the *structure* of the proof that is important here. If we want to prove something else about the function, we can use exactly the same structure. The proof assumes that we intuitively understand what is meant by the "number of nodes" in a binary tree. The recursively defined function NUMBER can be seen as a specification of this intuition.   ◆

Q Perfect binary trees. The set of **perfect binary trees** over A can be inductively defined in the following way:

- The empty tree, $(\,)$, is a perfect binary tree over A.
- If $x \in A$, and L and R are perfect binary trees over A *of the same height*, then (L, x, R) is a perfect binary tree over A.   ◆

The smallest perfect trees have the following shape, where each • represents a node in the binary tree:

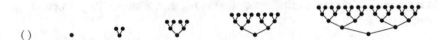

$(\,)$

❷ Show that the number of nodes in a perfect binary tree of height n is $2^n - 1$.   ◆

❶ We prove the statement by structural induction on the set of perfect binary trees.

The *base case* is that the statement holds for the smallest perfect binary tree, the empty tree $(\,)$. This has height equal to 0 and number of nodes equal to $2^0 - 1 = 0$, so the statement holds.

The *inductive step*: Suppose that the statement holds for two perfect binary trees L and R of the same height, and suppose that this height equals n. That is, the number of nodes in each of L and R equals $2^n - 1$. This is the *induction hypothesis*. We now get that (L, x, R) is a perfect binary tree of height $n + 1$. The total number of nodes in (L, x, R) is

$$\text{NUMBER}((L, x, R))$$
$$= \text{NUMBER}(L) + 1 + \text{NUMBER}(R) \qquad \text{by part (2) of the definition of NUMBER}$$
$$= (2^n - 1) + 1 + (2^n - 1) \qquad\qquad\qquad \text{by the induction hypothesis}$$
$$= 2(2^n - 1) + 1 \qquad\qquad\qquad\qquad\qquad \text{by ordinary calculation}$$
$$= 2^{n+1} - 1 \qquad\qquad\qquad\qquad\qquad\qquad \text{by ordinary calculation}$$

Thus, we have shown that the statement holds for the perfect binary tree (L, x, R).

By structural induction it follows that the statement holds for all perfect binary trees, that is, that the number of nodes in a perfect binary tree of height n is $2^n - 1$.   ◆

Some comments:

- Note that the answer in this exercise is exactly the same as in the analysis of Tower of Hanoi in Chapter 11 *(page 130)*. There we found that the least number of moves needed to move n disks from one rod to another is $2^n - 1$.
- We could have proved the statement by mathematical induction. Then the formulation would have been somewhat different, but the essence of the proof would have been exactly the same.

Exercises

12.1 Let f be a function on bit strings defined recursively in the following way:

(1) $f(0) = 1$ and $f(1) = 0$,

(2) $f(b0) = f(b)1$, where b is a bit string,

(3) $f(b1) = f(b)0$, where b is a bit string.

(a) What does this function do?

(b) What is $f(100)$?

(c) The following is a proof by induction that $f(f(b)) = b$ for all bit strings b. Determine what should be in the boxes.

The *base case*: This is that [1 _____] holds for $b = 0$ and $b = 1$. By substituting 0 for b, we get $f(f(0)) = 0$. The following calculation shows that this is true:

$$f(f(0)) = f(1) \qquad \text{by part (1) of the definition of f}$$
$$= 0 \qquad \text{by part (1) of the definition of f}$$

By substituting 1 for b, we get $f(f(1)) = 1$. The following calculation shows that this is true:

$$f(f(1)) = f(0) \qquad \text{by part (1) of the definition of f}$$
$$= 1 \qquad \text{by part (1) of the definition of f}$$

The [2 _____] *step*: Suppose that the statement holds for a bit string b, that is, that $f(f(b)) = b$. This is the [3 _____], and from this we must show that [4 _____] holds for a bit string bx, that is, that $f(f(bx)) = $ [5], where x is either 0 or 1. We get two cases, one for $x = 0$ and one for $x = 1$.

– If $x = 0$, we get the following:

$$f(f(b0)) = f(f(b)1) \qquad \text{by [6 ____] of the definition of f}$$
$$= f(f(b))0 \qquad \text{by part (3) of the definition of f}$$
$$= b0 \qquad \text{by the induction hypothesis}$$

– If $x = 1$, we get the following:

$$f(f(b1)) = f([7 ____]) \qquad \text{by [8 ____] of the definition of f}$$
$$= f(f(b))1 \qquad \text{by part (2) of the definition of f}$$
$$= b1 \qquad \text{by the induction hypothesis}$$

By [9 ____] *it follows that* [10 _____] *for all bit strings b.*

12.2 Let g be a function from bit strings to the integers defined recursively in the following way:

  – $g(0) = 1$ and $g(1) = -1$,
  – $g(b0) = g(b) + 1$,
  – $g(b1) = g(b) - 1$.

  (a) What does this function do?
  (b) What is $g(100)$?
  (c) Prove by induction that $g(b) + g(f(b)) = 0$ for all bit strings b, where f is the function defined in the previous exercise.

12.3 Let f be a function on $\{a, b, \square\}^*$ defined recursively by

  – $f(\Lambda) = \Lambda$,
  – $f(sx) = f(s)\square$, where s is a string and x is a character.

  (a) What is $f(aba)$?
  (b) What does this function do?
  (c) Prove by induction that $f(s) = f(f(s))$ for all $s \in \{a, b, \square\}^*$.

12.4 Let B be the language over the alphabet $\{(,)\}$ defined inductively like this:

  – $\Lambda \in B$,
  – if $X \in B$ and $Y \in B$, then $XY \in B$,
  – if $X \in B$, then $(X) \in B$.

  (a) Explain in your own words what the elements of the set B look like.
  (b) Prove by structural induction that for all $X \in B$, the number of characters in X is an even number.
  (c) Recursively define two functions, v and h, such that for all $X \in B$, we have that $v(X)$ is the number of left parentheses in X and $h(X)$ is the number of right parentheses in X.
  (d) Prove by structural induction that for all $X \in B$, the number of left parentheses in X equals the number of right parentheses in X.
  (e) Prove by structural induction that for all $X \in B$, if X is read from left to right, there is no position where more right parentheses have been read than left parentheses.

12.5 Let S be a set of propositional variables, and let v_1 and v_2 be valuations such that $v_1(X) = v_2(X)$ for all $X \in S$. Show by structural induction on the set of propositional formulas that

$$v_1(F) = v_2(F)$$

for all propositional formulas F that contain only propositional variables from S.

12.6 Let the function ANTI on the set of all propositional formulas be defined in the following way, where P stands for an arbitrary propositional variable and F and G stand for arbitrary propositional formulas:

$$\text{ANTI}(P) = \neg P,$$
$$\text{ANTI}(\neg F) = \neg\, \text{ANTI}(F),$$
$$\text{ANTI}(F \wedge G) = (\text{ANTI}(F) \vee \text{ANTI}(G)),$$
$$\text{ANTI}(F \vee G) = (\text{ANTI}(F) \wedge \text{ANTI}(G)),$$
$$\text{ANTI}(F \rightarrow G) = (\neg \text{ANTI}(F) \wedge \text{ANTI}(G)).$$

(a) Calculate ANTI$(P \vee \neg P)$ and ANTI$(\neg P \rightarrow Q)$.

(b) Explain in your own words what the function does.

(c) Prove by structural induction that ANTI(ANTI(F)) is equivalent to F, for all propositional formulas F.

(d) Is it the case that ANTI(ANTI(F)) = F for all propositional formulas F?

12.7 Let the functions s and u on the set of all propositional formulas be defined in the following way, where P stands for an arbitrary propositional variable and F and G stand for arbitrary propositional formulas:

$$s(P) = P, \qquad\qquad u(P) = \neg P,$$
$$s(\neg F) = u(F), \qquad\qquad u(\neg F) = s(F),$$
$$s(F \wedge G) = (s(F) \wedge s(G)), \qquad u(F \wedge G) = (u(F) \vee u(G)),$$
$$s(F \vee G) = (s(F) \vee s(G)), \qquad u(F \vee G) = (u(F) \wedge u(G)),$$
$$s(F \rightarrow G) = (u(F) \vee s(G)), \qquad u(F \rightarrow G) = (s(F) \wedge u(G)).$$

(a) Notice that the function s is defined in terms of u and vice versa. This is called **mutual recursion**. Explain why this is *not* a problem.

(b) Calculate $s(\neg(P \wedge Q))$ and $s(\neg\neg\neg P)$.

(c) Examine the functions and explain in your own words what they do.

(d) Prove by structural induction that F is equivalent to s(F) and that $\neg F$ is equivalent to u(F) for all propositional formulas F.

(e) Prove that s(F) is equivalent to u(u(F)) without using any kind of induction. Hint: Use (d).

(f) Prove by structural induction that s(F) = u(u(F)) for all propositional formulas F.

12.8 (a) Recursively define a function REV on lists that **reverses** the list. For example, REV$((1, 2, 3)) = (3, 2, 1)$. Hint: Use +.

(b) Check that REV$((a, b) + (c, d)) =$ REV$((c, d)) +$ REV$((a, b))$.

(c) Prove by structural induction on lists that

$$\text{REV}(L + M) = \text{REV}(M) + \text{REV}(L)$$

for all lists L and M.

(d) Prove by structural induction on lists that

$$\text{REV}(\text{REV}(L)) = L$$

for all lists L.

12.9   Recursively define a function MIRROR on binary trees that produces the mirror image of a tree:

$$\text{MIRROR}(\text{\Psi}) = \text{\Psi} \qquad \text{MIRROR}(\text{\it{J}}) = \text{\it{L}} \qquad \text{MIRROR}(\text{\Psi}) = \text{\Psi} \qquad \text{MIRROR}(\text{\Psi}) = \text{\Psi}$$

Use the inductive definition of binary trees when you define the function. Prove by structural induction that $\text{MIRROR}(\text{MIRROR}(T)) = T$ for all binary trees T.

12.10   Recursively define a function SUM on lists over numbers that sums all the numbers of the list. For example, $\text{SUM}((2,3)) = 5$ and $\text{SUM}((2,3,4)) = 9$. Prove by structural induction that $\text{SUM}(L + M) = \text{SUM}(L) + \text{SUM}(M)$ for all lists L and M.

12.11   Let f be a function. In this exercise we will look at the interplay between the recursively defined functions LEN and MAP on lists, which we encountered in Chapter 10 (*pages 117 and 118*). Prove by structural induction on lists that

$$\text{LEN}(\text{MAP}(f, L)) = \text{LEN}(L)$$

for all lists L, in other words, that applying MAP to a list does not change its length.

12.12   Let f be function. In this exercise we will look at the interplay between the recursively defined functions + and MAP on lists, which we encountered in Chapter 10 (*pages 117 and 118*). Prove by structural induction on lists that

$$\text{MAP}(f, (L + M)) = \text{MAP}(f, L) + \text{MAP}(f, M)$$

for all lists L and M, in other words, that applying f to the elements of a list is the same as applying f to the elements of the lists of which it is composed.

12.13   The set $\mathcal{O}$ that we defined in Chapter 9 (*page 108*) has many special properties. One of these is that if $X \in \mathcal{O}$, then $X \subseteq \mathcal{O}$. Prove this by structural induction.

Chapter 13

First-Order Languages

■■■■■■■■■■■■▮■■■■■■■■■■■■

In this chapter you will learn about the syntax of first-order logic. You will learn about first-order languages; logical and nonlogical symbols; constant, function, and relation symbols; as well as signatures, terms, formulas, and precedence rules.

Languages with Greater Expressibility

In propositional logic we have only propositional variables and the connectives ¬, ∧, ∨, and →. With these we can analyze and represent *propositions*. First-order logic, also called *predicate logic*, extends propositional logic with, among other things, *quantifiers*:

– ∃, the **existential quantifier**,
– ∀, the **universal quantifier**.

Using these, we can analyze and represent *quantified* propositions, that is, propositions expressing that an object with a particular property *exists* or that *all* objects have a particular property. Such statements are difficult to represent in propositional logic.

– *Each integer is either an even or an odd number.*
– *There are infinitely many prime numbers.*
– *Between every two rational numbers there is another rational number.*
– *Every even number greater than 2 is the sum of two prime numbers.*

Such *universal* and *existential* propositions often play a part in our reasoning. The following two arguments have different *content*, but exactly the same *form*.

All humans are mortal.	All perfect numbers are even numbers.
Kurt is not mortal.	x is not an even number.
Kurt is not a human.	x is not a perfect number.

Notice that both arguments are valid in terms of their form and not their content, and we can analyze such reasoning using first-order logic.

© Springer Nature Switzerland AG 2021
R. Antonsen, *Logical Methods*, https://doi.org/10.1007/978-3-030-63777-4_14

The plan for the next four chapters is to proceed as we did with propositional logic, just with a much richer language. It is therefore important that we be comfortable with all the concepts of propositional logic. We first define what a first-order *language* is and look at how a language gives rise to first-order *terms* and *formulas*. We then look at how the first-order formulas can be used to represent quantified propositions. Then we define first-order *models* and how terms and formulas can be interpreted in a unique way in a model. Finally, we look at some important concepts within first-order logic.

We distinguish sharply between *syntax*, which consists of formulas, symbols, and characters, and *semantics*, which consists of interpretations, truth values, and models. Syntax basically refers to what makes an arrangement of characters valid, irrespective of meaning, while semantics tells us how such an arrangement should be interpreted. We begin by defining the syntax of first-order logic.

First-Order Languages and Signatures

Here is an overview of what follows: To define first-order formulas, we must look at the smallest elements in the language and what symbols are allowed. We specify this in a so-called *signature*, which says which symbols are part of the language. A signature consists of constant symbols, function symbols, and relation symbols. When we come to the semantics, these will be interpreted respectively as elements, functions, and relations. A signature gives rise to an inductively defined set of first-order *terms*. The set of first-order *formulas* is then inductively defined in terms of the set of first-order terms. But all this depends on the underlying first-order language, and therefore we will start with that.

A first-order language can be divided into *logical* and *nonlogical* symbols. The *logical* symbols are those that are always interpreted in the same way, for example the connectives, and the *nonlogical* symbols are those that can be interpreted freely, in the same way as propositional variables in propositional logic.

Definition 13.1. First-order language

A **first-order language** consists of the following **logical symbols**: the logical connectives, $\wedge$, $\vee$, $\rightarrow$, and $\neg$, the **quantifiers**, $\forall$ and $\exists$, and an infinite countable set of **variables**, together with parentheses and commas. A first-order language also consists of the following **nonlogical symbols**: a set of **constant symbols**, a set of **function symbols**, and a set of **relation symbols**. The set of variables and the sets of constant, function, and relation symbols must be disjoint. Each function symbol and relation symbol is associated with a natural number, called the **arity** of the symbol.

The quantifiers ∀ and ∃ are read as "for all" and "there exists," respectively. The reason why the set of variables is infinite is that we want to have enough variables available, and the reason it is countable is that we want to have control over the set. A concrete way of defining the arities of function and relation symbols is by defining a function from the set of function and relation symbols to the natural numbers. The only thing that distinguishes two first-order languages from each other is the nonlogical symbols, and we capture this in the following definition.

Definition 13.2. Signature

The nonlogical symbols constitute what is called a **signature**. A signature is given as a tuple of three sets in the following way:

$$\langle \underbrace{a, b, c, \ldots}_{\text{constant symbols}} \; ; \; \underbrace{f, g, h, \ldots}_{\text{function symbols}} \; ; \; \underbrace{R, S, T, \ldots}_{\text{relation symbols}} \rangle$$

Here the constant, function, and relation symbols are separated by semicolons.

A signature is just a specification of what kind of nonlogical symbols there are in a first-order language. When we come to the semantics, it is the interpretation of these that determine which atomic formulas are true. When we specify a first-order language, it is sufficient to provide the signature, since the logical symbols do not change from one language to another.

Q **Three signatures and three languages.** Here are three signatures that we will refer to in several examples going forward. Each signature specifies its own first-order language.

- The signature $\langle a \, ; f, g \, ; P, R \rangle$ describes a simple first-order language in which a is a constant symbol, f and g are function symbols, and P and R are relation symbols; f and P are each of arity one, and g and R are each of arity two.
- The signature $\langle 0 \, ; s, + \, ; = \rangle$ describes a first-order language for number theory in which 0 is a constant symbol, s and $+$ are function symbols, and $=$ is a relation symbol; s has arity one, and $+$ and $=$ each have arity two. The symbol s comes from the word "successor," $+$ is the usual symbol that we use for addition, and $=$ is the usual symbol that we use for equality.
- The signature $\langle \varnothing \, ; \cap, \cup \, ; =, \in \rangle$ describes a first-order language for set theory in which $\varnothing$ is a constant symbol, $\cap$ and $\cup$ are function symbols, and $=$ and $\in$ are relation symbols. Each of the function and relation symbols is of arity two. ♦

First-Order Terms

The smallest components of each first-order language are first-order *terms*, and they are built up from variables, constant symbols, and function symbols.

Definition 13.3. First-order terms

Suppose that a first-order language is given. The set of **first-order terms**, or just **terms**, is inductively defined as the smallest set such that each variable and constant symbol is a **term**, and if f is a function symbol with arity n and $t_1, \ldots, t_n$ are terms, then $f(t_1, \ldots, t_n)$ is a **term**.

Q **Terms in the simple language.** The simplest terms in the language given by the signature $\langle a \, ; f, g \, ; P, R \rangle$ are the constant symbol a and all the variables x, y, z, etc. If t is a term, then $f(t)$ must also be a term, and we thus get that $f(a)$ and $f(x)$ are also terms. Then also $f(f(a))$ and $f(f(x))$ are terms. In the same way we have that if s and t are terms, then $g(s, t)$ is also a term, and we thus get that $g(a, a)$, $g(a, x)$, $g(x, a)$, and $g(x, x)$ are terms. By combining function symbols we get terms like $f(g(a, a))$, $f(g(a, x))$, $f(g(x, a))$, $g(f(a), a)$, $g(f(x), a)$, and $g(f(a), f(a))$. In the same way as with other inductively defined sets we get infinitely many distinct terms, but only a finite number of symbols in each individual term. ♦

Notation. As long as an expression is unambiguous and the arities of the symbols are known, we may drop the parentheses and, for example, write fa instead of $f(a)$, fx instead of $f(x)$, gaa instead of $g(a, a)$, and $gfaa$ instead of $g(f(a), a)$. But sometimes parentheses should be used to make a term more readable. For example, the term $gagafa$ becomes easier to read if we write $g(a, g(a, fa))$.

Q **Terms in the number theory language.** The simplest terms in the language given by the signature $\langle 0 \, ; s, + \, ; = \rangle$ are the constant symbol 0 and all the variables. Using s, we get terms like $s0$, $ss0$, and $sss0$. We can think of $sss0$ as a way to represent the number 3. Using $+$, we get terms like $+(x, y)$, $+(0, 0)$, $+(s0, 0)$, and $+(0, s0)$. But $++$, which consists of two successive function symbols, and $+(0)$, in which one argument is missing, are *not* admissible terms. ♦

Prefix, Infix, and Postfix Notation

If a function symbol with arity two is given, for example $+$, there are several alternative ways in which to write down composite terms:

- **prefix notation** is to place the function symbol at the *front*: $+(x, y)$
- **postfix notation** is to place the function symbol at the *back*: $(x, y)+$
- **infix notation** is to place the function symbol in the *middle*: $(x + y)$

In arithmetic and set theory, for example, it is natural to use infix notation. In set theory we do not write $\cap(A, B)$ with prefix notation, but $(A \cap B)$ with infix notation.

Q **Terms in the set theory language.** The simplest terms in the language given by the signature $\langle \varnothing\, ; \cap, \cup\, ; \, =, \in \rangle$ are the constant symbol $\varnothing$ and all the variables. Using $\cap$ and $\cup$, we get terms like $(x \cap y)$, $(\varnothing \cup z)$, and $((x \cap y) \cup z)$. ♦

Notice that no relation symbols appear in first-order terms. The intuition is that terms refer to elements of a set.

> **Digression**
>
> The postfix notation is also referred to as **Reverse Polish Notation** (RPN), after the Polish logician *Jan Łukasiewicz* (1878–1956), who introduced a similar notation at the end of the 1920s. In order to add 2 and 3, we must write 2 3 + in postfix notation. The advantage of this is that we do not need parentheses and we can accomplish more with fewer keystrokes. This notation has a rich history and may be used today in the **calc**-program in the text editor **Emacs**.

First-Order Formulas

We have now laid the foundation for defining first-order formulas. When a first-order *language* is given, the set of first-order *terms* is given. From these we can build up first-order *formulas* in the same way as in propositional logic: we begin with the *atomic* formulas and build up the *composite* formulas from these.

Definition 13.4. Atomic formula

If R is a relation symbol with arity $n > 0$ and $t_1, \ldots, t_n$ are terms, then $R(t_1, \ldots, t_n)$ is an **atomic formula**. If $n = 0$, then R is an **atomic formula**.

Notation. As long as an expression is unambiguous and the arities of the symbols are known, we may drop the parentheses and, for example, write Pfx, $Rafx$, and $Rgxaa$ instead of $P(f(x))$, $R(a, f(x))$, and $R(g(x, a), a)$.

The intuition is that atomic formulas are like sentences that can be either true or false, depending on the values the variables get, in contrast to terms, which are like the objects being referred to. Notice that we get a special situation when a relation symbol has arity zero. The definition gives an atomic formula without parentheses, and this will behave exactly like a propositional variable in propositional logic. Thus, we see that first-order formulas contain all propositional formulas, but also much more.

Q **Atomic formulas in the simple language.** Suppose again that the signature is $\langle a\, ; f, g\, ; P, R \rangle$. Here are some atomic formulas in this language:

$$Pa,\; Px,\; Pfx,\; Pffa,\; Pgaa,\; Raa,\; Rafx,\; Rgxaa\,.$$

For example, here $\mathsf{R}afx$ stands for $\mathsf{R}(a, \mathsf{f}(x))$, and this is unambiguous, because we know the arity of both R and f. The expression $\mathsf{P}(a, a)$ is not an atomic formula, because P has arity one, and $\mathsf{P}(\mathsf{R}(a, a))$ is not an atomic formula, because $\mathsf{R}(a, a)$ is not a term. ◆

Q **Atomic formulas in the number theory language.** Suppose again that the signature is $\langle 0 ; s, + ; = \rangle$. Here are some atomic formulas in this language:

$$0 = 0, \ s0 = 0, \ s0 + s0 = ss0, \ x = 0, \ x + y = y + x.$$

Try to see what the formulas are expressing. Notice that we do not have parentheses after s and that we are using infix notation for both $+$ and $=$. The expression $x = (y = 0)$ is not an atomic formula, because $(y = 0)$ is not a term, and $x + sx$ is not an atomic formula, because it is a term. ◆

Q **Atomic formulas in the set theory language.** Suppose we have the signature $\langle \varnothing ; \cap, \cup ; =, \in \rangle$. Here are some atomic formulas in this language:

$$x \in y, \ x \in (y \cap z), \ x = y, \ (x \cup y) = (x \cap y).$$

Here we are using infix notation for all the function and relation symbols. Try to see what the formulas express. ◆

We have finally reached the central definition of first-order formulas. The definition is inductive and assumes that the language, the terms, and the atomic formulas are already defined. The base set is the set of all atomic formulas, and new formulas are constructed from old, using connectives, quantifiers, and variables.

Definition 13.5. First-order formulas

Assume that a first-order language is given. The set of **first-order formulas**, or just **formulas**, is the smallest set such that:

- All atomic formulas are **formulas**.
- If φ and ψ are formulas, then $\neg\varphi$, $(\varphi \wedge \psi)$, $(\varphi \vee \psi)$, and $(\varphi \rightarrow \psi)$ are **formulas**.
- If φ is a formula and x is a variable, then $\forall x\varphi$ and $\exists x\varphi$ are **formulas**.

Each occurrence of a variable x in φ is said to be **bound** in the formulas $\forall x\varphi$ and $\exists x\varphi$ and within the **scope** of the associated quantifier.

Some comments are in place:

- A first-order language gives rise to exactly one set of first-order formulas.
- The definition says nothing about what formulas *mean*. We have defined only the *syntax* of first-order formulas, that is, what comprises the admissible first-order formulas. We have intuitions about how the formulas should be interpreted, but this does not become precise until we have defined the *semantics*.

– The letters φ and ψ in the definition are the Greek letters *phi* and *psi*. Because they are used only as placeholders in the definition, they could equally well have been other letters, for example F and G. But there are many who use these Greek letters for first-order formulas, and so shall we.

– The concept of *scope* is the same as what we find in programming languages. When we declare or define a variable in a programming language, it also has a specific scope, and that is where the variable is visible. How this is defined varies from one programming language to another.

Slowly, but surely, we will build up our intuition about what first-order formulas express before we define the semantics. Therefore, in the following examples, there are some natural ways to read the formulas.

Q **Composite formulas in the simple language.** Here are some composite first-order formulas in the simple language, together with what they express.

Pa	P *is true for* a
$\exists x Px$	*there exists an* x *such that* Px
$\forall x Px$	*for all (or for every)* x *it is such that* Px
$\neg Rax$	a *is not related to* x
$\forall x \forall y Rxy$	*the relation is universal / everything is related to everything else*
$\exists x \exists y Rxy$	*the relation is not empty / something is related to something else*
$\neg \exists x \exists y Rxy$	*the relation is empty / nothing is related*
$\exists x \forall y Rxy$	*there is an element that is related to every element*
$\forall x \exists y Rxy$	*for every element there is an element to which it is related*
$\exists x Rxx$	*there exists an element related to itself*
$\forall x (Px \to Rxx)$	*if* P *is true for an element, then it is related to itself*
$\forall x (Rxx \to Px)$	P *is true for every element that is related to itself*

We can thus formulate some familiar properties of relations:

$\forall x Rxx$	R *is reflexive*
$\forall x \forall y (Rxy \to Ryx)$	R *is symmetric*
$\forall x \forall y \forall z (Rxy \land Ryz \to Rxz)$	R *is transitive*
$\forall x \neg Rxx$	R *is irreflexive*
$\neg \forall x Rxx$	R *is not reflexive*
$\forall x \forall y (Rxy \to \neg Ryx)$	R *is asymmetric*

Antisymmetry cannot be described in this language, because the language does not contain the equality symbol, $=$. ♦

Q **Composite formulas in the number theory language.** Here are some composite first-order formulas in the number theory language together with what they express:

$\neg(s0 = 0)$	*it is not the case that one equals zero*
$\forall x \forall y (x + y = y + x)$	$x + y$ *always equals* $y + x$
$\exists x (s0 + s0 = x)$	*there is a number that equals one plus one*
$\neg \exists x (sx = 0)$	*zero is not the successor of any number*
$\forall x \forall y (sx = sy \rightarrow x = y)$	*a number has only one predecessor / s is injective*
$\forall x (x + 0 = x)$	x *plus zero is always* x

For example, this language contains neither $\cdot$ nor $<$, and therefore $0 < s0$ and $\forall x (x \cdot 0 = 0)$ are not formulas in this language. ◆

Q **Composite formulas in the set theory language.** Here are some composite first-order formulas in the set theory language together with what they express:

$\forall x (x \in y \rightarrow x \in z)$	y *is a subset of* z
$\forall x (x \cup \varnothing = x)$	*to take the union with* $\varnothing$ *has no effect*
$\forall x (x \cap \varnothing = \varnothing)$	*to intersect with* $\varnothing$ *always gives* $\varnothing$
$\forall x (x \in x)$	*every element is an element of itself*
$\forall x \forall y \forall z (x \cap (y \cup z) = (x \cap y) \cup (x \cap z))$	$\cap$ *distributes over* $\cup$

In this language, for example, we do not have the function symbol $\setminus$ or the relation symbol $\subseteq$. ◆

Precedence Rules

Just as for propositional formulas, we have precedence rules for first-order formulas. The set of first-order formulas is precisely defined, but we will also accept, for example, $\exists y Py \wedge Px$ as a first-order formula, although the parentheses are missing. We give the connectives different **precedences** in relation to each other just as we did for propositional formulas in Chapter 2 (*page 25*). We say that $\forall$ and $\exists$ *bind the strongest* and are equal in strength to $\neg$. For example, $\exists y Py \wedge Px$ stands for $(\exists y Py \wedge Px)$, where the scope of the existential quantifier is just Py, and not $\exists y (Py \wedge Px)$. Everything else is the same as what was done for propositional logic. Notice that parentheses are not introduced for $\neg$, $\forall$, or $\exists$ in the definition of first-order formulas. This is because they are not needed.

Exercises

13.1   Find first-order formulas in the language $\langle Ola, Kari \,;\; ; Mother, Father \rangle$ for the following sentences. The relation symbols have arity two.

(a) *Ola is the father of Kari.*

(b) *Kari is the mother of someone.*

(c) *Ola has no mother.*

(d) *Everyone has a mother and a father.*

(e) *Everyone has a grandmother.*

(f) *No one is both a mother and a father.*

13.2   Decide which of the following expressions are first-order *terms*, given that the signature is $\langle a, b \,; f, g \,; P, R \rangle$, f and P each have arity one, and g and R each have arity two. If the expression is not a *term*, explain why not.

(a) x

(b) $f(x)$

(c) $g(x, b)$

(d) $g(a)$

(e) $f(x, x)$

(f) $P(x) \wedge R(x, a)$

(g) $\forall x P x$

(h) $f(f)$

(i) $f(P(a))$

(j) $f(f(f(f(f(f(f(x))))))))$

13.3   Decide which of the following expressions are first-order *formulas*, given that the signature is $\langle a, b \,; f, g \,; P, R \rangle$, f and P each have arity one, and g and R each have arity two. If the expression is not a formula, explain why not.

(a) $R(x, b)$

(b) $\forall x R(x, b)$

(c) $R(\forall x P x, b)$

(d) $P \vee R(x, x)$

(e) $\exists \forall x P x$

(f) $R(f(f(a)), w)$

(g) $\exists y \wedge \forall y P y$

(h) $P x \rightarrow \exists y R(y)$

(i) $\forall x P(x, a)$

(j) $\forall y R(a)$

(k) $\exists z P(a)$

(l) $\exists z \exists y \forall z P(z)$

(m) $(P x)$

(n) $f(P(x))$

(o) $R(P(a), b)$

(p) $(\exists x P x)$

(q) $\forall x f(x)$

(r) $P x \exists x x$

13.4   Use the precedence rules to place all the parentheses in the right place:

(a) $\neg \quad \forall x \quad \neg \quad P x \quad \rightarrow \quad Q x \quad \wedge \quad R x$

(b) $\exists x \quad P x \quad \wedge \quad Q x \quad \vee \quad \exists y \quad \neg \quad R y$

(c) $P x \quad \rightarrow \quad Q x \quad \wedge \quad \forall x \quad R x \quad \rightarrow \quad S x$

(d) $\forall x \quad \exists y \quad P x \quad \wedge \quad Q y \quad \rightarrow \quad \neg \quad R y$

(e) $\forall x \quad \exists y \quad \forall z \quad \exists u \quad \forall v \quad \exists w \quad \neg \quad R x$

13.5   In the inductive definition of first-order formulas, parentheses are introduced only for the connectives $\wedge$, $\vee$, and $\rightarrow$. Why are parentheses not introduced for $\neg$, $\forall$, and $\exists$?

13.6   Why does the definition of precedence rules not cover which of $\neg$, $\forall$, and $\exists$ binds the strongest?

13.7   What is the difference between $\exists x Px \wedge Qx$ and $\exists x(Px \wedge Qx)$? Do the precedence rules tell you how the expressions should be read in this case?

13.8   What is the difference between a function symbol and a function? What is the difference between a relationship symbol and a relation?

13.9   We usually use infix notation in ordinary mathematics, but there are also symbols that use prefix and postfix notation. Which?

13.10   Why do you think it is a requirement that the set of variables and the sets of constant, function, and relation symbols must be disjoint?

13.11   Give three examples of propositions that cannot be represented in a good way in propositional logic.

13.12   What is the reason we distinguish between logical and nonlogical symbols?

13.13   Are there always infinitely many terms in a first-order language?

13.14   Are there always infinitely many formulas in a first-order language?

13.15   Can a first-order language have uncountably many terms or formulas?

Chapter 14

Representation of Quantified Statements

█■■■■■■■■■■■■┊■■■■■■■■■■

In this chapter you will learn about predicates, properties related to free variables, and how first-order languages can be used to represent quantified propositions.

Representation of Predicates

We can use first-order languages to represent propositions in much greater detail than what is possible with propositional logic. We already know a lot from propositional logic: we can both read propositional formulas in the right way and use propositional formulas to represent propositions.

- *Is it the case that the statement "x + y is an even number" is true?*
- *It is impossible to answer this, because the statement contains two placeholders.*
- *OK, let's assume x equals 5.*
- *But that doesn't help. We are left with the statement "5 + y is an even number," and the truth value depends on what y is.*

To move on, it is useful to have the concept of a *predicate*.

Definition 14.1. Predicate

A **predicate** is an expression containing one or more placeholders that becomes true or false when we replace the placeholders with values.

Q **Predicate.** The expression "∘ *is an even number*" is a predicate. If we replace the placeholder ∘ with 5, we get something that is false. The expression "∘ *likes* •" is a predicate, but with two placeholders. It will be either true or false when ∘ and • are replaced with values. ♦

We know predicates well from proofs by induction, where it usually is a statement with one or more placeholders that is being proved for all elements of an inductively defined set.

We are going to represent predicates as atomic formulas in first-order languages, where variables are used to indicate the different placeholders. For example, "x

© Springer Nature Switzerland AG 2021
R. Antonsen, *Logical Methods*, https://doi.org/10.1007/978-3-030-63777-4_15

is a good person" is a predicate that can be represented by the atomic first-order formula Gx. Here G is a relation symbol with arity one. It makes sense that it has arity one because the predicate that is represented has one placeholder. When we have represented predicates as atomic formulas, we can add quantifiers and obtain *composite* first-order formulas. Now, for example, we can represent the proposition "there is at least one good person" with the formula ∃xGx, and the proposition "all persons are good" with the formula ∀xGx. The following overview shows different ways of reading such expressions:

∃xGx	∀xGx
There is an x *such that* Gx	*For all* x, *it is such that* Gx
There is an x *such that* x *is a good person*	*For all* x, x *is a good person*
There exists a good person	*Everyone is a good person*

There is nothing wrong with adding a quantifier with another variable, but it is quite pointless. The formula ∃yGx can be read as "there is a y such that x is good." In this proposition, x still serves as a placeholder, and the whole expression can therefore still be interpreted as a predicate, even though there is a quantifier outside. We say that x is a *free* variable in the formula ∃yGx.

Syntactic Properties of Free Variables

There is a fundamental difference between the formulas Px and ∀xPx. In the formula Px, the variable x is not within the scope of a quantifier, but it is in ∀xPx. This is such a useful term that we capture it in a definition.

> **Definition 14.2. Free variables and closed formulas**
>
> A variable occurrence in a first-order formula is **free** if it is not **bound**, that is, if it is not within the **scope** of a quantifier. A formula is **closed** if it does not contain any free variables.

Q **Free/bound variables and closed formulas.** In the formula ∀xRxy ∧ Pz, we have that x is bound, while both y and z are free. In the formula ∀xPxy → ∀zPzx, we have that x is both bound and free, y is free, and z is bound. None of these formulas are closed, but the formula ∀xPxa is closed, because it does not contain any free variables. The formula Pab is also *closed* because it does not contain any free variables.

♦

The distinction between closed and nonclosed formulas is useful. Previously, we defined a *proposition* as something that is either true or false. A closed formula is in practice a *proposition*. When we get to the semantics, we will see how *closed formulas* can be interpreted so that they get a truth value, true or false.

The Art of Expressing Yourself with a First-Order Language

We will now define a first-order language for representing propositions about *admiration*, with the signature $\langle$ a, b ; ; *Idol, Likes* $\rangle$. That is, a and b are constant symbols, there are no function symbols, and the relation symbols are *Idol* and *Likes*. We assume that *Idol* has arity one and that *Likes* has arity two. The following are examples of expressions that are represented with *atomic* formulas:

- *Idol*(a) represents *Alice is an idol.*
- *Likes*(a, b) represents *Alice likes Bob.*

  It is important that the representations be made systematically. Here we decide that *Likes*(x, y) represents "x likes y," and we then have to stick with it. We could have represented "Alice likes Bob" with *Likes*(b, a), with the order of a and b interchanged, but then we must assume that *Likes*(x, y) represents "y likes x," and translate the propositions accordingly.
- *Likes*(x, a) represents x *likes Alice.*

  Here the expression "x likes Alice" is a predicate, and it will not have any truth value until a value for x is specified. In a similar way we see that the formula *Likes*(x, a) is not *closed*, because x is a free variable.

The following are examples of expressions that are represented with *composite* formulas. Notice that all the formulas are closed.

- $\forall$x*Likes*(a, x) represents *Alice likes everyone.*

  For all x, Alice likes x.
- $\exists$x*Idol*(x) represents *There is an idol.*

  There is an x such that x is an idol.
- $\forall$x(*Likes*(b, x) $\rightarrow$ *Likes*(a, x)) represents *Alice likes everyone that Bob likes.*

  For all x, if Bob likes x, then Alice likes x. Here the parentheses are important: the *whole* expression following the quantifier is within parentheses.
- $\exists$x*Likes*(x, x) represents *Someone likes themself.*

  There is an x such that x likes x.
- $\forall$x(*Likes*(x, x) $\rightarrow$ *Likes*(b, x)) represents *Bob likes everyone who likes themself.*

  For all x, if x likes x, then Bob likes x.
- $\neg\exists$x(*Likes*(x, a) $\wedge$ *Likes*(x, b)) represents *No one likes both Alice and Bob.*

  It is not such that there is an x such that x likes Alice and x likes Bob. Here it is also possible to write $\forall$x$\neg$(*Likes*(x, a) $\wedge$ *Likes*(x, b)); for all x, it is *not* such that x likes Alice and x likes Bob.
- $\exists$x$\neg$*Likes*(x, x) represents *Someone does not like themself.*

  There is an x such that x does not like x. It is also possible to say $\neg\forall$x*Likes*(x, x). This means that it is *not* the case that everyone likes themself.
- $\exists$x(*Likes*(b, x) $\wedge$ *Likes*(x, a)) represents *Bob likes someone who likes Alice.*

  There is an x such that Bob likes x and such that this x likes Alice. We can work from within and outward in the following way: That an x likes Alice can be represented by *Likes*(x, a). Furthermore, that Bob likes this x can be represented by *Likes*(b, x).

– $\forall x \exists y Likes(x, y)$ represents *Everyone likes someone.*

For every x, there is a y such that x likes y.

– $\forall x(\forall y Likes(y, x) \rightarrow Idol(x))$ represents *Someone who is liked by everyone is an idol.*

For all x, if everyone likes this x, then x is an idol. That everyone likes x can be represented by $\forall y Likes(y, x)$.

– $\forall x(Idol(x) \rightarrow \forall y Likes(y, x))$ represents *An idol is liked by everyone.*

For all x, if x is an idol, then x is liked by everyone.

Choice of First-Order Language

Different languages are used for different purposes. Sometimes we want a rich language with many symbols to express many things in a compact way; at other times, we want a small language in order to be in control and do a lot with simple means. There is no given solution here, and the signature we choose depends on what we want to do.

In the previous chapter, we saw a simple language for number theory with the signature $\langle 0; s, +; = \rangle$. In this language, a natural number was represented by the term $sss\ldots0$, where there were as many occurrences of s as the number we are representing. Instead of doing it this way, we can, for example, include one constant symbol for each individual natural number, and we then get infinitely many constant symbols, or we can add 1 as a new constant symbol. In both cases we can skip s as a function symbol. If we add 1, we can represent a natural number as $1 + 1 + \cdots + 1$, with as many occurrences of 1 as the number we are representing. There is no one correct approach, and we must choose the signature that suits us best. In this language there is also no function symbol for multiplication or a relation symbol for "less than." In the next example we look at a language in which these symbols are in the signature.

Q   **Another language for number theory:** $\langle 0, 1; +, \cdot; =, < \rangle$. Here we no longer have the function symbol s, but instead we have the constant symbol 1, the function symbol $\cdot$, and the relation symbol $<$. We assume that all function and relation symbols have arity two. The simplest terms in this language are the constant symbols 0 and 1, and all the variables. Using $+$ and $\cdot$ we get terms like $1 + 1, 0 \cdot 1, x + 0, x \cdot x$, and $(1 + 1) \cdot (1 + 1)$. Here we are using infix notation for both $+$ and $\cdot$. Here are some atomic formulas in this language, together with what they express:

$x < 1$	*x is less than 1*
$x = x$	*x equals x*
$x \cdot 1 = x$	*x multiplied by 1 equals x*
$(1 + 0) < (1 + 1)$	*1 + 0 is less than 1 + 1*

Here are some composite formulas in the language, together with what they express:

$\neg(x < x)$	*x is not less than itself*
$\forall x(x < x + 1)$	*x is always less than x + 1*
$\forall x(x \cdot 0 = 0)$	*x multiplied by 0 is always 0*
$\forall x \exists y(x < y)$	*for every number there is a number that is larger*
$\neg\exists y \forall x(x < y)$	*there is no largest number*
$\forall x(0 < x)$	*0 is the smallest number*
$\forall x \forall y \forall z(x \cdot (y + z) = (x \cdot y + x \cdot z))$	*· distributes over +*

And so we can continue. We have not yet said anything about whether the formulas are true. All of these formulas can actually be made true or false depending on the interpretation that we use. ♦

Repeating Patterns in Representations

There are some patterns that recur when we represent propositions with first-order logic:

– $\forall x(\varphi \to \psi)$ is often used to represent propositions of the form "every φ is ψ." A common mistake is to write $\forall x(\varphi \wedge \psi)$. Do not write this unless it is exactly what you want to say. It means that "everything is both φ and ψ."

– $\exists x(\varphi \wedge \psi)$ is often used to represent propositions of the form "there is something that is both φ and ψ." A common mistake to write $\exists x(\varphi \to \psi)$. Do not write this. It means that "there is something such that if it is φ, then it is ψ." It is a strange thing to say, and probably not what you want to say.

– $\neg\exists x\varphi$ and $\forall x\neg\varphi$ mean the same thing, and both are used to represent propositions of the form "there is not something such that φ" and "nothing is φ." A common mistake is to write $\exists x\neg\varphi$ or $\neg\forall x\varphi$. Do not write this unless it is exactly what you want to say. It means that "there is something that is not φ" and "not everything is φ."

Repetition of First-Order Languages

The definition of first-order formulas is quite extensive, and it may be instructive to see all the parts together. Suppose that a first-order language is given. *The set of first-order terms is inductively defined*: each variable and each constant symbol is a first-order term, and if f is a function symbol with arity n and $t_1, \ldots, t_n$ are terms, then $f(t_1, \ldots, t_n)$ is a term. *The set of first-order formulas is also inductively defined*: If R is a relation symbol with arity n, and $t_1, \ldots, t_n$ are terms, then $R(t_1, \ldots, t_n)$ is an atomic formula. If φ and ψ are formulas, then $\neg\varphi$, $(\varphi \wedge \psi)$, $(\varphi \vee \psi)$, and $(\varphi \to \psi)$ are formulas. If φ is a formula and x is a variable, then $\forall x\varphi$ and $\exists x\varphi$ are formulas.

A variable x is *bound* if it is within *the scope* of a quantifier. A variable occurrence is *free* if it is not bound. A term or formula that does not contain free variables is said to be *closed*. Parentheses are added only for $\wedge$, $\vee$, and $\to$. These are the only places where they are needed to eliminate ambiguities.

Expressibility and Complexity

First-order logic is richer than propositional logic, and with first-order languages we are able to express far more complex connections than with only propositional variables and connectives.

For propositional logic there are methods to *decide* whether formulas are valid, and for all practical purposes these methods are fast. By adding the quantifiers in first-order logic, the expressibility and complexity increase so much that we can *prove* that such algorithms no longer exist. To decide whether a first-order formula is valid is so difficult that there is no general algorithm that will always, regardless of the formula, answer yes or no correctly after finitely many steps. We say that first-order logic is *undecidable*.

It is nevertheless worth noting that we are quantifying only over variables, which intuitively represent elements of a set. We are quantifying over neither constant nor function nor relation symbols. If we had done that, we would have gained even greater expressibility.

Digression

Second-order logic is an extension of first-order logic whereby in addition to quantifying over variables that represent elements, we quantify over variables that represent functions and relations. This naturally gives an even greater expressibility. For example, in second-order logic we may express that a set is finite, something we cannot express in first-order logic. The way to do this is by saying that every injective function on a set is also surjective. This is a property that holds only for finite sets; if a set is infinite, there exists a function on it that is injective but not surjective. It is also possible to define equality in second-order logic by saying that $x = y$ is true if and only if $\forall R(Rx \leftrightarrow Ry)$, that is, if a property holds for x, it also holds for y, and vice versa. If we limit second-order logic in different ways, we get more manageable logics. If we quantify only over sets, that is, over unary relations, we get so-called **monadic second-order logic**, which also has many interesting properties.

Exercises

14.1 Decide whether each of the following expressions can be interpreted as a predicate. If so, specify what the placeholders are and give examples of replacements of the placeholders with values such that the expression becomes true and such that it becomes false: (a) *x is an even number.* (b) *The house is 100 years old.* (c) *If F is true, then G is true.* (d) *Without drink and food, the hero is no good.*

14.2 Suppose that P, Q, R, and S are relation symbols such that Px represents "x is popular"; Qx, "x is controversial"; Rxy, "x is friends with y"; and Sxy, "x has voted for y." Suppose that a, b, and c represent Anna, Bernt, and Carl, respectively. Find first-order formulas for the following sentences: (a) *Anna is popular and controversial.* (b) *Bernt is friends with Carl, but not with Anna.* (c) *Carl is friends with everyone.* (d) *Anna voted for someone who is controversial.* (e) *Bernt voted for a friend.* (f) *Carl is friends with someone who is popular.* (g) *There is someone who is not popular.* (h) *No one is both popular and controversial.* (i) *No one votes for Anna.* (j) *Everyone has voted for someone who is controversial.* (k) *Everyone who is controversial has been voted for by someone.* (l) *No one voted for themself.* (m) *There is someone who has not been voted for by anyone.* (n) *Everyone who is popular is friends with everyone.* (o) *Anna is friends with everyone who is popular.* (p) *Anna is friends only with those who are popular.* (q) *Bernt votes only for controversial and unpopular friends of his.* (r) *Carl is not voting for anybody.*

14.3 Find first-order formulas for the following sentences. Choose your own first-order language. (a) ***All** children are kind.* (b) ***No** children are kind.* (c) ***Some** children are kind.* (d) ***All** children who are kind get a reward.* (e) ***Some** children who are kind get a reward.* (f) ***Every** child is cute or not cute.* (g) ***All** provable formulas are valid.* (h) ***There are two** sheriffs in town.*

14.4 For each of the following formulas, find out which variables that occur are free and which are bound:

(a) Rxy

(b) $\forall x Sxy$

(c) $\forall x \exists y Txy$

(d) $\forall x Rxy \land Px$

(e) $\forall x (Rxy \land Px)$

(f) $\forall y (Ryy \land Py)$

(g) $\exists x (y + x = 0)$

(h) $\forall x \forall y (y + x \rightarrow x + y)$

(i) $\forall x \exists y (y + x = 0)$

14.5 Suppose that M, T, and L are relation symbols such that Mx represents "x is a mathematician," Tx that "x is a TV celebrity," and Lxy that "x likes y." Find *natural-sounding* English sentences that correspond to the following first-order formulas:

(a) $\neg \exists x (Mx \land Tx)$

(b) $\exists x (Tx \land \neg Mx)$

(c) $\exists x \forall y (Ty \rightarrow Lxy)$

(d) $\neg \exists x Mx$

(e) $\forall x (Mx \lor Tx)$

(f) $\forall x (Tx \rightarrow Lxx)$

(g) $\exists x (Tx \land \exists y (My \land Lyx))$

(h) $\forall x \exists y Lxy$

(i) $\forall x \exists y (Lxy \land \neg Ty)$

(j) $\exists x (Mx \land \forall y Lxy)$

(k) $\forall x (Mx \rightarrow \forall y (Ty \rightarrow \neg Lxy))$

(l) $\neg \exists x \exists y (Mx \land Ty \land Lxy)$

14.6 Suppose that B, F, and K are relation symbols such that Bx represents "x is a biologist," Fx represents "x is a philosopher," and Kxy represents "x knows y." Suppose that a, b, and c are constant symbols that represent Aristotle, Bolzano, and Copernicus, respectively. Find first-order formulas for the following sentences: (a) *Aristotle is both a biologist and a philosopher.* (b) *All biologists are philosophers.* (c) *No philosophers are biologists.* (d) *Aristotle knows a philosopher.* (e) *Bolzano knows all philosophers.* (f) *Copernicus knows only biologists.* (g) *Everyone knows a philosopher.* (h) *Everyone knows someone who knows a philosopher.* Find *natural-sounding* English sentences that correspond to the following first-order formulas:

(i) $\neg \exists x (Bx \lor Fx)$

(j) $\exists x (Kax \land \neg Bx)$

(k) $\forall x \forall y ((Fx \land Fy) \to Kxy)$

(l) $\exists x \forall y (Fy \to Kxy)$

14.7 Find first-order formulas for the following sentences. Choose an appropriate language. (a) *There is a ball, and if this ball is red, I like it.* (b) *Only the cool ones come to the party.* (c) *All the cool ones come to the party.* (d) *No one drinking a gallon of coffee in the evening falls asleep.* (e) *If the relation R is symmetric, it is also transitive.*

14.8 Suppose that S, E, F, and L are relation symbols such that Sx represents "x is a student," Exy represents "x loves y," Fxy represents "x is in love with y," and Lxy represents "x likes y." Suppose that a, b, and c are constant symbols that represent Anna, Bernt, and Carl, respectively. Find first-order formulas for the following sentences: (a) *Everyone loves Bernt.* (b) *Anna loves Bernt, but Bernt does not love Anna.* (c) *Carl loves all students.* (d) *Carl loves only students.* (e) *Some students are in love with Carl.* (f) *Bernt loves Anna and does not like those who are in love with her.* (g) *Bernt is in love with a student who is in love with Carl.* (h) *Anyone who loves a person both likes and is in love with that person.* (i) *There is someone who is not in love.* (j) *Someone is in love with someone they do not like.* (k) *Someone is in love with someone who is not in love with them.* (l) *Anna does not like those who like everyone.* (m) *Bernt likes only those who love themselves.*

14.9 Suppose that B, P, S, Q, and T are relation symbols such that Bx represents "x reads good books," Px represents "x is a professor," Sx represents "x is a student," Qxy represents "x is smarter than y," and Txy represents "x watches more TV than y." Suppose that a, b, and c are constant symbols that represent Anna, Bernt, and Carl, respectively. Find first-order formulas for the following sentences: (a) *Anna and Bernt are both professors who do not read good books.* (b) *There is no professor who watches more TV than a student.* (c) *Carl is smarter than all students who watch more TV than Anna.*

Find *natural-sounding* English sentences that correspond to the following first-order formulas:

(d) $\forall x (Sx \to Bx)$

(e) $\forall x (Bx \to \exists y (Qxy \land \neg By))$

Chapter 15

Interpretation in Models

■■■■■■■■■■■■■■■■|■■■■■■■■■

In this chapter you will learn about the semantics of first-order logic: what models are and how they are used to interpret first-order terms and formulas. You will learn about how substitutions are used in the interpretation of quantified formulas, and about the terms valid, satisfiable, contradictory, and falsifiable.

Semantics for First-Order Logic

To impose a semantics on a language is to say something about the relationship between language and reality. In propositional logic, it is the *valuations* that provide the semantics. In first-order logic, it is the *models* that do the corresponding job. We are going to look at how first-order formulas are interpreted and what makes first-order formulas true and false. This will enable us to study what makes formulas valid, satisfiable, contradictory, falsifiable, equivalent, and, perhaps most important of all, what makes a first-order formula a logical consequence of other first-order formulas.

A model intuitively consists of a set called the *domain* of the model and an interpretation of all nonlogical symbols such that

- a constant symbol is interpreted as an *element* of the set;
- a function symbol is interpreted as a *function* on the set;
- a relation symbol is interpreted as a *relation* on the set.

We will soon define more precisely what a model is and how models provide truth values for first-order formulas, but first we will look at some examples to get an intuitive grip on what is going on.

Q **Truth depends on interpretation.** Look at the formula ∀xPx. We read it as **"for all** x, Px is true." Whether this formula is *true* or *false* depends on how the relation symbol P is *interpreted* in a model. If P is interpreted as the set of all elements in the domain, the formula is true. If not, it is false. Look at the formula Px. We read it as "Px is true." The formula contains a free variable, and whether it is true or false depends on what happens with x. This is the same situation as with the formula $x + 2 = 5$. Whether it is true in a model depends on what happens with x. ♦

© Springer Nature Switzerland AG 2021
R. Antonsen, *Logical Methods*, https://doi.org/10.1007/978-3-030-63777-4_16

Q   **Truth depends on interpretation.** Look at the formula $\forall x \exists y (x < y)$. We read it as **"for every** x, **there exists** a y such that $x < y$." Whether this formula is true or false depends on how the relation symbol $<$ is *interpreted* in a model. If $<$ is interpreted as the less-than relation over the natural numbers, it is true. The reason is that for every natural number, there is a natural number that is larger than it. If $<$ is interpreted as the less-than relation over the set $\{1, 2, 3, 4, 5, 6, 7\}$, it is false. The reason is that there is a number, namely 7, in relation to which there is no larger number.   ◆

Definition of Model

To specify a model for a language intuitively means specifying a domain and stating how all the nonlogical symbols are to be interpreted. We then use a model to interpret first-order terms and formulas. If, for example, the language consists of only one relation symbol, R, it suffices to specify the domain, which is a set, and the interpretation of R, which is a relation on that set. If the domain is the natural numbers and R has arity two, the interpretation of R must be a binary relation on the natural numbers.

Definition 15.1. Model

A **model** $\mathcal{M}$ for a given first-order language consists of a nonempty set D, called the **domain** of the model, and a function $\cdot^{\mathcal{M}}$ (written as a superscript) that interprets all nonlogical symbols in the following way:

– If k is a constant symbol, then $k^{\mathcal{M}} \in D$.
– If f is a function symbol with arity n, then $f^{\mathcal{M}}$ is a function from D^n to D.
– If R is a relation symbol with arity n, then $R^{\mathcal{M}}$ is an n-ary relation on D, that is, a subset of D^n.

We write $|\mathcal{M}|$ for the domain D of the model $\mathcal{M}$.

Think of a model as a kind of machine that can answer questions. If you ask it about what its domain is, you get a nonempty set in response. If you ask it about what a symbol means, you get an interpretation of that symbol in response: a constant symbol gives you an element of the domain in response, a function symbol gives you a function in response, and a relation symbol gives you a relation in response.

– Writing the name of the model "$\mathcal{M}$" after a symbol is a practical and widely used notation. We can read $\circ^{\mathcal{M}}$ as "$\circ$ interpreted in $\mathcal{M}$" no matter what $\circ$ stands for. This notation makes it completely clear that we are talking about the *interpretation* of a symbol and not the symbol itself.
– The domain of a model cannot be empty, and that is for good reason. We could have allowed empty domains, but then we would have had another semantics.

Interpretation of Terms

We are going to interpret first-order terms as elements of a set. The terms belong to the syntax, the elements belong to the semantics, and the models tell us how the terms are to be interpreted, in other words, which elements they represent. To make it as simple as possible, we are interpreting only terms that do not contain variables.

Definition 15.2. Closed term

A term is **closed** if it does not contain any variables.

The following definition tells us how to interpret an arbitrary closed term in a model.

Definition 15.3. Interpretation of terms

Let a first-order language be given, and let $\mathcal{M}$ be a model for this language. We then interpret a closed term $f(t_1, \ldots, t_n)$ in the following way:

$$f(t_1, \ldots, t_n)^{\mathcal{M}} = f^{\mathcal{M}}\left(t_1^{\mathcal{M}}, \ldots, t_n^{\mathcal{M}}\right).$$

This is an example of a *recursive* function. The model $\mathcal{M}$ gives an interpretation of constant and function symbols, and this makes up the base case for the recursion. We then extend interpretations to apply to composite terms by saying that the term $f(t_1, \ldots, t_n)$ interpreted in $\mathcal{M}$ equals what we get by applying the function $f^{\mathcal{M}}$ to the result of interpreting the terms $t_1, \ldots, t_n$ in $\mathcal{M}$. Here is an example that shows how terms are interpreted in a model.

Q **Interpretation of terms.** Let $\mathcal{M}$ be a model with domain $\{0, 1, 2, 3, 4\}$ such that:

- The constant symbol a is interpreted as 3, and we write this as $a^{\mathcal{M}} = 3$.
- The constant symbol b is interpreted as 4, and we write this as $b^{\mathcal{M}} = 4$.
- The function symbol f with arity one is interpreted as a function on $\{0, 1, 2, 3, 4\}$ that gives 0 if the argument is an even number and 1 if the argument is an odd number.

Now we can interpret all terms that contain these symbols. A model provides a systematic way of interpreting all terms in a language.

- The interpretation of $f(a)$ is written as $f(a)^{\mathcal{M}}$. By calculating inward we get $f(a)^{\mathcal{M}} = f^{\mathcal{M}}(a^{\mathcal{M}}) = f^{\mathcal{M}}(3) = 1$.
- The interpretation of $f(b)$ is written as $f(b)^{\mathcal{M}}$. By calculating inward we get $f(b)^{\mathcal{M}} = f^{\mathcal{M}}(b^{\mathcal{M}}) = f^{\mathcal{M}}(4) = 0$. ◆

Interpretation of Atomic Formulas

We have seen how closed terms can be interpreted in a model, and we can proceed to define the interpretation of atomic formulas. What we do here is an extension of propositional logic, and we take with us all the concepts from there. In propositional logic we first assigned truth values, **0** or **1**, to propositional variables, and then we defined truth values for all propositional formulas. We are going to do the same thing now. First, we shall give truth values to atomic formulas. Then we will define truth values for first-order formulas.

To find out whether a *closed* atomic formula is true in a model, all relation symbols and terms in the formula must be interpreted. Then we can check whether the interpretation of the relation symbols is in accordance with the interpretation of the terms.

Definition 15.4. Interpretation of a closed atomic formula

Suppose that $\mathcal{M}$ is a model for a first-order language, and let $R(t_1, \ldots, t_n)$ be a closed atomic formula. We say that the formula $R(t_1, \ldots, t_n)$ is **true** in $\mathcal{M}$ and write $\mathcal{M} \models R(t_1, \ldots, t_n)$ if the following holds:

$$\langle t_1^{\mathcal{M}}, \ldots, t_n^{\mathcal{M}} \rangle \in R^{\mathcal{M}}.$$

To interpret a closed atomic formula $R(t_1, \ldots, t_n)$, we must first interpret the relation symbol R and all the terms $t_1, \ldots, t_n$ individually. Then we get a relation and a number of elements. By checking whether the tuple consisting of the elements is an element of the relation, we find out whether the atomic formula is true.

Q   **Interpretation of an atomic formula.** Suppose $\mathcal{M}$ is a model with the natural numbers as domain, that $a^{\mathcal{M}} = 3$, $b^{\mathcal{M}} = 4$, and that $R^{\mathcal{M}}$ is the less-than relation on the natural numbers. Then Rab is true in $\mathcal{M}$, because $a^{\mathcal{M}}$, which equals 3, is less than $b^{\mathcal{M}}$, which equals 4. Here there is a correspondence between the closed atomic formula Rab, which is purely syntactic, and the binary relation that constitutes the interpretation of R. We say that Rab is true in the model $\mathcal{M}$ because $\langle a^{\mathcal{M}}, b^{\mathcal{M}} \rangle \in R^{\mathcal{M}}$. ♦

Q   **A simple model.** Suppose we are talking about books and that we have a language that consists of the relation symbols N and E, where Nx represents the predicate "the book x is written in Norwegian" and Ex represents the predicate "the book x is written in English." Suppose the model $\mathcal{M}$ has a set of books as domain, that N is interpreted as the Norwegian books, and that E is interpreted as the English books. We can write $N^{\mathcal{M}} =$ the set of Norwegian books, and $E^{\mathcal{M}} =$ the set of English books. Suppose further that a is a constant symbol in the language. If a is interpreted as a Norwegian book, the formula Na becomes true in the model. This is because $a^{\mathcal{M}} \in N^{\mathcal{M}}$. And if a is interpreted as an English book, the formula Ea becomes true in the model, and this is because $a^{\mathcal{M}} \in E^{\mathcal{M}}$. ♦

Substitutions

In order to give the definition of how formulas are interpreted in a model, we need the concept of *substitution*. The reason why substitutions are introduced right now is that they are used to interpret formulas with quantifiers. As a preliminary intuition, we shall say that $\forall x Px$ is true in a model if Px is true no matter what we substitute for x. We need the concept of substitution to make what we mean by "substituting for x" completely precise. We define substitution first in terms, then in formulas.

Definition 15.5. Substitution in terms

Let s and t be terms, and x a variable. Then $s[x/t]$ is the result of replacing, or **substituting**, all occurrences of x in s with t.

We can view $[x/t]$ as a function that replaces every x with a t. One way to make this completely explicit is by recursively defining a substitution function.

Q Substitution in terms. If we replace every occurrence of x with a in the term $f(x, y)$, we get $f(a, y)$. We write this as $f(x, y)[x/a] = f(a, y)$. Here are some more examples, where the colors only indicate what is replaced with what:

$$f(x, y, a)[x/y] = f(y, y, a)$$
$$f(y, y, a)[y/b] = f(b, b, a)$$

$$(x + y)[x/3] = 3 + y,$$
$$(x + y)[y/3] = x + 3.$$
♦

Definition 15.6. Substitution in formulas

If φ is a formula, t is a term, and x is a variable, then $\varphi[x/t]$ is the result of replacing, or **substituting**, all *free* occurrences of x in φ with t.

Substitution in formulas is a bit more complex than substitution in terms. The reason is that if a variable is bound by a quantifier, the substitution has no effect on that variable. We can imagine that the quantifier protects the variable. In the formula $\exists x(x + y = 100)$, the x occurs bound, and substituting x with a number should have no effect; the result should just be the formula itself. The variable y, on the other hand, occurs free, and it may freely be substituted with a number in the following way:

$$\exists x(x + y = 100)[y/25] = \exists x(x + 25 = 100).$$

Q Substitution in formulas.

$$(Pxy \wedge \forall xPxy)[x/a] = (Pay \wedge \forall xPxy),$$
$$(Pxy \wedge \forall xPxy)[y/a] = (Pxa \wedge \forall xPxa).$$
♦

Interpretation of Composite Formulas

We are almost ready to interpret all first-order formulas. In previous examples, we have often had constant symbols for all the elements of a domain. It is useful, but not always the case, that these constant symbols are in the language. Strictly speaking, we have no guarantee that a language is so rich. For simplicity – and to define the semantics – we will assume that for each model and language, the language contains constant symbols for all elements of the domain of the model.

Assumption. If $\mathcal{M}$ is a model for a given first-order language, we assume that for each element a in $|\mathcal{M}|$, there is a constant symbol $\bar{a}$ in the language. We assume that each model $\mathcal{M}$ interprets $\bar{a}$ as a, in other words, that $\bar{a}^{\mathcal{M}} = a$.

This assumption plays an important role in the interpretation of first-order formulas. We have managed fine without it thus far, but we need it now to avoid the following definition becoming less elegant and more technically demanding.

Intuitively, $\exists$- and $\forall$-formulas will be interpreted in the following way: a formula $\exists x \varphi$ is true if we can insert something for x in φ such that φ becomes true. For example, the formula $\exists x(x - 5 = 0)$ is true, because the expression $(x - 5 = 0)$ becomes true when we insert 5 for x. Similarly, a formula $\forall x \varphi$ is true if φ becomes true no matter what we insert for x. For example, the formula $\forall x(x - x = 0)$ is true, because $(x - x = 0)$ becomes true no matter what we insert for x. The following definition makes this precise.

Definition 15.7. Interpretation of closed formulas

Suppose that $\mathcal{M}$ is a model for a given first-order language. We recursively define what it means for a closed formula φ to be **true** in $\mathcal{M}$. The notation $\mathcal{M} \models \varphi$ means that φ is true in the model $\mathcal{M}$. The base case, for atomic formulas, is already defined. This says that $\mathcal{M} \models R(t_1, \ldots, t_n)$ if $\langle t_1^{\mathcal{M}}, \ldots, t_n^{\mathcal{M}} \rangle \in R^{\mathcal{M}}$. Here are the recursive steps:

$$\mathcal{M} \models \neg \varphi \qquad \text{if it is } \textit{not} \text{ the case that } \mathcal{M} \models \varphi.$$
$$\mathcal{M} \models \varphi \wedge \psi \qquad \text{if } \mathcal{M} \models \varphi \textit{ and } \mathcal{M} \models \psi.$$
$$\mathcal{M} \models \varphi \vee \psi \qquad \text{if } \mathcal{M} \models \varphi \textit{ or } \mathcal{M} \models \psi.$$
$$\mathcal{M} \models \varphi \rightarrow \psi \qquad \text{if } \mathcal{M} \models \varphi \textit{ implies } \mathcal{M} \models \psi.$$
$$\mathcal{M} \models \forall x \varphi \qquad \text{if } \mathcal{M} \models \varphi[x/\bar{a}] \textit{ for all } a \text{ in } |\mathcal{M}|.$$
$$\mathcal{M} \models \exists x \varphi \qquad \text{if } \mathcal{M} \models \varphi[x/\bar{a}] \textit{ for at least one } a \text{ in } |\mathcal{M}|.$$

When φ is true in $\mathcal{M}$, we also say that $\mathcal{M}$ is a **model for** φ, that $\mathcal{M}$ makes φ true, and that $\mathcal{M}$ **satisfies** φ. All of these mean the same thing.

Notice that substitutions are used in the last two items of the definition. The reason that x is replaced with $\bar{a}$, and not with a, is that $\bar{a}$ is the constant symbol that represents the element a. The element a belongs to the semantics, and $\bar{a}$ belongs to the syntax; when we replace x, which is a character, we must replace it with something that is also a character. It is exactly because we have assumed that we have constant symbols for all the elements of the domain of each model that we can say that *for all* a in $|\mathcal{M}|$, the formula $\varphi[x/\bar{a}]$ must be true in $\mathcal{M}$. The last item of the definition can be read as, "the model $\mathcal{M}$ makes $\exists x \varphi$ true if it makes $\varphi[x/\bar{a}]$ true for at least one element a of the domain of $\mathcal{M}$."

Q   **Language with other symbols.** The symbols to include in a language may be chosen completely freely. Suppose the signature is $\langle \text{⚑}, \text{⚑}, \text{⚘} ; \text{⚔} ; \text{♀}, \text{♂} \rangle$, and that $\mathcal{M}$ is a model for this language. Then $\text{⚑}^{\mathcal{M}}$, $\text{⚑}^{\mathcal{M}}$, and $\text{⚘}^{\mathcal{M}}$ must be elements of the domain, $\text{⚔}^{\mathcal{M}}$ must be a function on the domain, and $\text{♀}^{\mathcal{M}}$ and $\text{♂}^{\mathcal{M}}$ must be relations on the domain. We can assume that ⚔ has arity two and that ♀ and ♂ have arity one.

  – Some terms in this language are x, ⚑, $\text{⚔}(\text{⚑}, \text{⚑})$, and $\text{⚔}(\text{⚑}, \text{⚔}(\text{⚑}, \text{⚘}))$. The first term is pretty boring, since it is just a variable. The second term, ⚑, must be interpreted as an element of the domain of $\mathcal{M}$. The third term is also interpreted as an element of the domain, namely as the element that is obtained by applying a function, the interpretation of ⚔, to the elements that are the interpretations of ⚑ and ⚑.
  – Some atomic formulas in this language are $\text{♂}(\text{⚑})$, $\text{♀}(\text{⚑})$, and $\text{♀}(\text{⚔}(\text{⚑}, \text{⚑}))$. One way of interpreting the symbol ♂ is as the set of all men, and one way of interpreting the symbol ♀ is as the set of all women. In this case, the atomic formula $\text{♂}(\text{⚑})$ expresses that ⚑ is a man. Similarly, $\text{♀}(\text{⚔}(\text{⚑}, \text{⚑}))$ expresses that $\text{⚔}(\text{⚑}, \text{⚑})$ is a woman.
  – Some composite formulas in this language are $\neg \text{♂}(\text{⚑})$, $\exists x (\text{♂}(x) \wedge \text{♀}(x))$, and $\forall x (\text{♀}(x) \rightarrow \text{♂}(\text{⚔}(x, \text{⚘})))$. For example, the second formula is false when ♀ and ♂ are interpreted as disjoint sets.   ◆

Satisfiability and Validity of First-Order Formulas

The concepts *satisfiable*, *contradictory*, *valid*, and *falsifiable* in first-order logic are completely analogous to those we know from propositional logic. The only difference is that we are now quantifying over models instead of valuations. To justify that a formula is valid, we must say something about all models.

Definition 15.8. Satisfiable, contradictory, valid, and falsifiable

A closed formula is **satisfiable** if there is a model that makes it true; otherwise, it is **contradictory**. A closed formula is **valid** if all models makes it true; otherwise, it is **falsifiable**. A set of closed formulas is satisfiable/falsifiable if there is a model that makes all the formulas true/false.

The concept of *satisfiability* is important. If, for example, you have a scientific theory, it is interesting if it is satisfiable, in other words, if there exists a model that makes all of its formulas true. If it is impossible to make all the formulas true simultaneously, it is a pretty bad theory. A good scientific theory should be such that it is possible to make it true. Another example is *specification of computer programs*. If a program specification is given, it should be possible to write a program that meets the specification. This is what the concept of *satisfiability* is about.

In the same way, *falsifiability* is an equally important concept and an often used criterion for calling something scientific. If it is impossible to make all the formulas of a scientific theory *false* simultaneously, that is, to falsify the theory, it is in several ways an uninteresting theory, and many would say that it then does not say anything substantial about the world.

❓ Show that the formula $\forall x(Px \lor \neg Px)$ is valid.   ◆

❗ Let $\mathcal{M}$ be an arbitrary model, and let e be an arbitrary element of the domain of $\mathcal{M}$. No matter how P is interpreted, we have either $e \in P^{\mathcal{M}}$ or $e \notin P^{\mathcal{M}}$. In the first case, $\mathcal{M} \models P\bar{e}$. In the second case, $\mathcal{M} \models \neg P\bar{e}$. This means that $\mathcal{M} \models P\bar{e} \lor \neg P\bar{e}$ for all e, and this means that $\mathcal{M} \models \forall x(Px \lor \neg Px)$.   ◆

❓ Make a model that satisfies both $\exists x Px$ and $\forall x \neg Qx$.   ◆

❗ Let the domain of the model $\mathcal{M}$ be $\{1\}$, that is, let $|\mathcal{M}| = \{1\}$. There are no constant or function symbols in the language, and we therefore do not need to specify the interpretation of these. Let the relation symbols P and Q be interpreted such that $P^{\mathcal{M}} = \{1\}$ and $Q^{\mathcal{M}} = \varnothing$. The formula $\exists x Px$ is true because $P\bar{1}$ is true, and $P\bar{1}$ is true because $1 \in P^{\mathcal{M}}$. The formula $\forall x \neg Qx$ is true because $\neg Q\bar{1}$ is true, and $\neg Q\bar{1}$ is true because $Q\bar{1}$ is false, and $Q\bar{1}$ is false because $1 \notin Q^{\mathcal{M}}$.   ◆

❓ Make a model that satisfies the formulas $Pa \land Pb$, $\neg \exists x(Px \land Qx)$, and $\exists x Qx$.   ◆

❗ Let the domain of the model be $\{1, 2\}$, that is, $|\mathcal{M}| = \{1, 2\}$. Let the constant symbols be interpreted such that $a^{\mathcal{M}} = b^{\mathcal{M}} = 1$. There are no function symbols in the language. Interpret the relation symbols such that $P^{\mathcal{M}} = \{1\}$ and $Q^{\mathcal{M}} = \{2\}$. The formula $Pa \land Pb$ is true because $1 \in P^{\mathcal{M}}$. The formula $\neg \exists x(Px \land Qx)$ is true because $(P^{\mathcal{M}} \cap Q^{\mathcal{M}}) = \varnothing$. The formula $\exists x Qx$ is true because $2 \in Q^{\mathcal{M}}$.   ◆

❓ Show that the formula $(\forall x Px \land \forall x Qx) \rightarrow \forall x(Px \land Qx)$ is valid.   ◆

❗ The formula is of the form $\varphi \rightarrow \psi$, and we must show that $\mathcal{M} \models \varphi \rightarrow \psi$ for all $\mathcal{M}$. Let $\mathcal{M}$ be an arbitrary model, and let D be the domain of $\mathcal{M}$. To show that $\mathcal{M} \models \varphi \rightarrow \psi$, it is sufficient to show that if $\mathcal{M}$ makes φ true, then $\mathcal{M}$ also makes ψ true. Assume (A1) that $\mathcal{M}$ makes $(\forall x Px \land \forall x Qx)$ true. From assumption (A1) we must show that $\mathcal{M}$ makes $\forall x(Px \land Qx)$ true. Assume (A2) that a is an arbitrary element of the domain D. From assumption (A1) it follows that $\mathcal{M}$ makes both $\forall x Px$ and $\forall x Qx$ true. From this and assumption (A2) it follows that $P\bar{a}$ and $Q\bar{a}$ are both true in $\mathcal{M}$. Then $P\bar{a} \land Q\bar{a}$ must also be true in $\mathcal{M}$. Because a was arbitrarily chosen, it follows that $\forall x(Px \land Qx)$ is true in $\mathcal{M}$.   ◆

❷ Show that the formula $\forall x(Px \lor Qx) \to (\forall xPx \lor \forall xQx)$ is falsifiable.   ◆

❶ We must find a model that makes the formula false. So we must find a model $\mathcal{M}$ that makes $\forall x(Px \lor Qx)$ true, but $\forall xPx \lor \forall xQx$ false. Then the model $\mathcal{M}$ must make both $\forall xPx$ and $\forall xQx$ false. Let the domain of $\mathcal{M}$ be $\{1, 2\}$. In order to make $\forall xPx$ false, let $P^{\mathcal{M}} = \{1\}$. Then $P\bar{2}$, and therefore also $\forall xPx$, is false. In order to make $\forall xQx$ false, let $Q^{\mathcal{M}} = \{2\}$. Then $Q\bar{1}$, and therefore also $\forall xQx$, must be false. It is easy to check that $\forall x(Px \lor Qx)$ is true in $\mathcal{M}$.   ◆

❷ Show that the formula $\forall xRxx \to \forall x\exists yRxy$ is valid.   ◆

❶ Let $\mathcal{M}$ be a model with domain D, and suppose that $\mathcal{M} \models \forall xRxx$. It is sufficient to show that $\mathcal{M} \models \forall x\exists yRxy$. Let $a \in D$ be an arbitrarily chosen element. From the assumption we get that $\mathcal{M} \models R\bar{a}\bar{a}$. Then $\mathcal{M} \models \exists yR\bar{a}y$. Because a was arbitrarily chosen, $\mathcal{M} \models \forall x\exists yRxy$.   ◆

Digression

The Norwegian mathematician and logician *Thoralf Albert Skolem* (1887–1963) was one of the world's foremost logicians of all time. He wrote his doctoral thesis on number theory and algebra, and he made groundbreaking contributions in mathematical logic, especially within proof theory, recursion theory, and axiomatic set theory. Several theorems and concepts today bear Skolem's name. For example, we have the *Skolem–Löwenheim theorem* in set theory, which says that if a countable first-order theory has an infinite model, it also has a countable model. This has been called *Skolem's paradox*, because of the seemingly paradoxical claim that all theories, also theories for uncountable sets, have countable models. So-called Skolem functions are another example. In the formula $\forall x\exists yR(x, y)$ there is one existential quantifier, and we can remove it if we simultaneously replace y with an appropriate functional expression $f(x)$. We then get the formula $\forall xR(x, f(x))$, and this is satisfiable if and only if the original formula is satisfiable. The function symbol f is here called a **Skolem function**, and the process that removes the existential quantifier and inserts a functional expression is called **Skolemization**. This is a useful tool, for example, when we program automated theorem provers.

First-Order Languages and Equality

When we have the equality symbol $=$ in a first-order language, we have to pay attention. In principle, we are allowed to treat this symbol like any other relation symbol, but it is often useful to assume that every model interprets $=$ as the identity relation. If we do not make this assumption but we still want to interpret $=$ correctly, we must take care of this in another way, for example by adding formulas that *axiomatize* equality.

Assumption. If the relation symbol $=$ is included in a signature, we assume that it is a relation symbol with arity two and that each model interprets $=$ as the identity relation, that is, that $=^{\mathcal{M}}$ equals $\{\, \langle x, x \rangle \mid x \in |\mathcal{M}| \,\}$.

Q **First-order formulas with equality.** Here are some examples of properties we know and that can be represented in first-order languages with equality:

$\forall x \forall y (Rxy \wedge Ryx \to x = y)$ R *is antisymmetric*
$\forall x \forall y (f(x) = f(y) \to x = y)$ f *is injective*
$\forall y \exists x (f(x) = y)$ f *is surjective*
$\forall x (f(x) = x)$ f *is the identity function*

By means of equality we can also express that there are at least n elements of a domain. Here we use $s \neq t$ as an abbreviation for $\neg(s = t)$.

$\exists x (x = x)$ *there is at least one element*
$\exists x \exists y (x \neq y)$ *there are at least two elements*
$\exists x \exists y \exists z (x \neq y \wedge x \neq z \wedge y \neq z)$ *there are at least three elements*
$$\vdots$$

This means that we can easily make sure that every model that satisfies the formulas of a set has at least a given size for its domain. We just need to add one of these formulas to the set.   ◆

A Little Repetition

When we worked with functions and relations in Chapters 6 and 7, we did not make such a sharp distinction between syntax and semantics as we have done now. There we used the symbol R both as a *symbol* and as the relation it *represented*. Here we are a little more precise.

Models can be complex objects, but we most often use them as *interpretation functions*. We can look at the act of attaching a small $\mathcal{M}$ to a symbol as an instruction to *interpret* this symbol, and then the model behaves like a function: it takes a symbol as an argument and returns the interpretation of the symbol as its value. For example, if R is a relation symbol, then $R^{\mathcal{M}}$ is a relation.

Recall that the arity of function and relation symbols determines how they should be *interpreted*. If R is a relation symbol with arity 1, it must be interpreted as a *unary* relation; if it has arity 2, it must be interpreted as a *binary* relation.

We have learned that if a is an element of the domain of a model, then $\bar{a}$ is a *constant symbol* that *stands for* and *represents* the element a. We have assumed that all languages contain all such constant symbols and that these are interpreted correctly, that is, that $\bar{a}^{\mathcal{M}} = a$. This can be compared to *pointers* in programming languages. A pointer is something that *points to* and *stands for* something other than itself, and this is exactly how the bar over the letter is used. We can think of $\bar{a}$ as a *pointer* that points to a.

Exercises

15.1 Provide a *natural* model for each first-order language:

(a) $\langle 0; s, +; = \rangle$

(b) $\langle 0, 1; +, \times; =, < \rangle$

15.2 Show that the following formulas are valid:

(a) $\forall x(Px \lor Qx) \to (\exists xPx \lor Qb)$

(b) $(\forall xRxa \lor \forall xRxb) \to \forall x(Rxa \lor Rxb)$

Show that the following formulas are falsifiable:

(c) $(\exists xPx \land \exists xQx) \to \exists x(Px \land Qx)$

(d) $\forall x(Rxa \lor Rxb) \to \exists x(Rax \land Rbx)$

(e) $(\forall xRxa \lor \forall xRxb) \to \forall x(Rax \lor Rbx)$

(f) $\forall x(Px \to Qx) \to (\exists xPx \to \forall xQx)$

15.3 Suppose that $\mathcal{M}$ is a model for $\langle a, b, c; s, f, g; \rangle$. Suppose that $\mathcal{M}$ has the natural numbers as domain and is such that $a^{\mathcal{M}} = 1$, $b^{\mathcal{M}} = 2$, and $c^{\mathcal{M}} = 3$, and for all natural numbers x, $f^{\mathcal{M}}(x) = 2x$ and $g^{\mathcal{M}}(x, y) = (x \cdot y) + 1$. Calculate the following:

(a) $f(a)^{\mathcal{M}}$

(b) $f(b)^{\mathcal{M}}$

(c) $f(f(c))^{\mathcal{M}}$

(d) $g(a, c)^{\mathcal{M}}$

(e) $f(g(b, c))^{\mathcal{M}}$

(f) $g(g(a, a), g(c, c))^{\mathcal{M}}$

15.4 For each of the following expressions, determine whether it mainly represents something *syntactic* or *semantic*: x, $\forall xPx$, $\mathcal{M}$, $f^{\mathcal{M}}$, $\models$, $\overline{1089}$, *Small*, *Rab*, $\Leftrightarrow$, $A \subseteq B$, $|\mathcal{M}|$, $((2+3)+4) = 9$, *function*, *arity*, *relation symbol*, *validity*, *unary relation*, and *closed*.

15.5 Is it meaningful to write $\mathcal{M} \models \forall xRxy$? In that case, what does it mean? Hint: Read Definition 15.7 *(page 172)*.

15.6 Must all relation symbols be interpreted as nonempty relations or are empty relations allowed? Hint: Use the definition.

15.7 Are $\exists xPx$, $\forall x(Px \to Qx)$ and $\exists x\neg Qx$ satisfiable in one and the same model?

15.8 Explain briefly what is needed to make the following formulas *false*:

(a) $Pa \land Qa$

(b) $Pa \lor Qa$

(c) $Pa \to Qa$

(d) $\forall x(Px \to Qx)$

(e) $\exists x(Px \lor Qx)$

(f) $\exists x(Px \to Qx)$

15.9 Show that the formulas (a) $Pa \lor Pb \to \exists xPx$ and (b) $\forall xPx \to Pa \land Pb$ are valid.

15.10 For each of the following formulas, find a model that makes the formula true. Let the domain be $\{1, 2\}$. It is sufficient to interpret the relation symbol R.

(a) $\forall x \forall y Rxy$

(b) $\exists x \forall y Rxy$

(c) $\forall x \exists y Rxy \land \neg \exists xRxx$

(d) $\exists x \exists y (Rxy \land \neg Ryx) \land \forall xRxx$

15.11 What can you say about the truth value of the formula $\forall xPx \to \exists xPx$?

15.12 For each of the following formulas, determine whether the formula is valid:

(a) $\forall x \forall y (Rxy \rightarrow \neg Ryx) \rightarrow \forall x \neg Rxx$ (c) $\exists x (Px \rightarrow Pa \wedge Pb)$

(b) $\forall x \neg Rxx \rightarrow \forall x \forall y (Rxy \rightarrow \neg Ryx)$ (d) $\forall x (Px \rightarrow Pa \wedge Pb)$

15.13 Let $\mathcal{M}$ be a model with $\mathbb{N}$ as domain, and suppose the following:

$a^{\mathcal{M}} = 2$, $b^{\mathcal{M}} = 3$, and $c^{\mathcal{M}} = 4$, $g^{\mathcal{M}}(x, y) = x + y$,

$f^{\mathcal{M}}(x) = x + 2$, $R^{\mathcal{M}}(x, y)$ if x is strictly smaller than y.

(a) What is $g(f(a), g(b, c))^{\mathcal{M}}$?   (c) Is it true that $\mathcal{M} \models \exists x \forall y R(x, f(y))$?

(b) Is it true that $\mathcal{M} \models R(g(a, b), c)$?   (d) Is it true that $\mathcal{M} \models \forall x \exists y R(y, x)$?

15.14 For each of the following formulas, find a model that makes the formula *false*. Let the domain be $\{1, 2, 3\}$. It is sufficient to specify the interpretation of the relation symbol R.

(a) $\forall x \forall y Rxy$ (c) $\forall x \exists y Rxy \wedge \neg \exists x Rxx$

(b) $\exists x \forall y Rxy$ (d) $\exists x \exists y (Rxy \wedge \neg Ryx) \wedge \forall x Rxx$

15.15 (a) Define substitution of variables for terms recursively. That is, for two arbitrary terms s and t, recursively define what $s[x/t]$ is. For example, $f(x, a, y)[x/g(a)] = f(g(a), a, y)$.

(b) Define substitution of variables for formulas recursively. That is, for an arbitrary formula φ and an arbitrary term t, recursively define what $\varphi[x/t]$ is. Make sure that bound variables are not replaced. For example, $\forall x R(x, y)[x/g(a)] = \forall x R(x, y)$, but $\forall x R(x, y)[y/g(a)] = \forall x R(x, g(a))$.

15.16 Let the signature $\langle c \, ; f \, ; R \rangle$ be given, and suppose that R has arity two, and f arity one. Find a model $\mathcal{N}$ with domain $\{1, 2, 3\}$ that makes the formula $\forall x R(x, f(x))$ true. Then let $\mathcal{M}$ be the model with domain $\{0, 1, 2, 3\}$ such that:

– $c^{\mathcal{M}} = 0$

– $f^{\mathcal{M}} = \{ \langle 0, 1 \rangle, \langle 1, 2 \rangle, \langle 2, 3 \rangle, \langle 3, 0 \rangle \}$

– $R^{\mathcal{M}} = \{ \langle 0, 1 \rangle, \langle 0, 2 \rangle, \langle 0, 3 \rangle, \langle 1, 3 \rangle, \langle 2, 3 \rangle \}$

Determine which of the following first-order formulas are true in $\mathcal{M}$:

(a) $R(c, f(c))$ (d) $\forall x \exists y (f(y) = x)$

(b) $\forall x R(x, x)$ (e) $\forall x \forall y (R(x, y) \rightarrow R(f(x), f(y)))$

(c) $\forall x (x = c \vee R(c, x))$

15.17 Suppose we have a first-order language with two binary relation symbols, R and $=$. Let $\mathcal{M}$ be a model with domain $\mathcal{P}(U)$. Suppose that R and $=$ are interpreted as the subset relation $\subseteq$ and the equality relation. Decide whether the following formulas are true or false in the model $\mathcal{M}$. Provide short reasons for each answer.

(a) $\exists x (x = x)$ (c) $\exists x \forall y Rxy$

(b) $\forall x \forall y (x = y)$ (d) $\forall x \exists y (\neg (x = y) \wedge Rxy)$

(e) $\forall x \forall y \exists z \forall t (Rzx \wedge Rzy \wedge (Rtx \wedge Rty \rightarrow Rtz))$ Hint: Think about intersection.

Chapter 16

Reasoning About Models

■■■■■■■■■■■■■■■┊■■■■■■■■

*In this chapter you will learn more about models in first-order logic. You will
learn about logical equivalence and consequence, the interaction between
the quantifiers and the connectives, modeling, theories, and axiomatizations,
and a little bit about prenex normal form.*

Logical Equivalence and Logical Consequence

The concepts of *logical equivalence* and *logical consequence* in first-order logic correspond precisely to those we know from propositional logic.

Definition 16.1. Equivalence

Two closed first-order formulas φ and ψ are **equivalent** if each model that makes
φ true also makes ψ true, and vice versa. Put another way, for each model $\mathcal{M}$, we
have $\mathcal{M} \models \varphi$ if and only if $\mathcal{M} \models \psi$. We write $\varphi \Leftrightarrow \psi$ when φ and ψ are equivalent.

The formulas φ and ψ are equivalent if and only if the formula $(\varphi \to \psi) \wedge (\psi \to \varphi)$ is
valid. All valid formulas are equivalent to each other, and all contradictory formulas
are equivalent to each other.

Definition 16.2. Logical consequence

Let M be a set of closed first-order formulas, and let φ be a closed first-order formula.
If φ is true in each model that makes all the formulas in M true simultaneously,
then φ is a **logical consequence**, or just a **consequence**, of the formulas in M. We
write $M \models \varphi$ when φ is a logical consequence of M. By the notation $\varphi \Rightarrow \psi$ we
mean that ψ is a logical consequence of the set consisting of φ.

The formula ψ is a logical consequence of φ if and only if the formula $(\varphi \to \psi)$ is
valid. A valid formula is a logical consequence of all formulas, and all formulas are
logical consequences of a contradiction.

© Springer Nature Switzerland AG 2021
R. Antonsen, *Logical Methods*, https://doi.org/10.1007/978-3-030-63777-4_17

The Interaction Between Quantifiers and Connectives

Just as there is an interplay between negation and the other connectives, there is an interplay between negation and the quantifiers. The first relationships we look at are often called **De Morgan's laws,** corresponding to those we met in Chapter 3 *(page 36)* for propositional logic. These can be used to find equivalent formulas in which the negation symbols are as close to the relation symbols as possible.

Q **Quantifiers and negation.** Here are some examples of typical equivalences between first-order formulas and what they mean:

$\neg \forall x P x$ is equivalent to $\exists x \neg P x$
If not everyone comes to the party, someone is not coming, and vice versa.
$\neg \exists x P x$ is equivalent to $\forall x \neg P x$
If there is nothing dangerous, everything is nondangerous, and vice versa.

It is a good exercise to prove these. In order to prove the first one, we must prove the following:

($\Rightarrow$) For all models $\mathcal{M}$, if $\mathcal{M} \models \neg \forall x P x$, then $\mathcal{M} \models \exists x \neg P x$.
($\Leftarrow$) For all models $\mathcal{M}$, if $\mathcal{M} \models \exists x \neg P x$, then $\mathcal{M} \models \neg \forall x P x$.

Let $\mathcal{M}$ be an arbitrary model. We can see that both ($\Rightarrow$) and ($\Leftarrow$) hold by means of the following equivalences:

$\mathcal{M} \models \neg \forall x P x \quad \Leftrightarrow \quad$ it is not the case that $\mathcal{M} \models \forall x P x$
$\Leftrightarrow \quad$ it is not the case that for all elements $a \in |\mathcal{M}|$, $\mathcal{M} \models P\bar{a}$
$\Leftrightarrow \quad$ there is an element $a \in |\mathcal{M}|$ such that not $\mathcal{M} \models P\bar{a}$
$\Leftrightarrow \quad$ there is an element $a \in |\mathcal{M}|$ such that $\mathcal{M} \models \neg P\bar{a}$
$\Leftrightarrow \quad \mathcal{M} \models \exists x \neg P x$.

From these it follows that $\neg \forall x \neg P x$ is equivalent to $\exists x P x$ and that $\neg \exists x \neg P x$ is equivalent to $\forall x P x$. This means that as long as we have negation, we can express everything by means of a single quantifier.   ♦

Just as there is an interplay between the connectives in propositional logic, there is an interplay between the quantifiers and the connectives: the quantifiers *distribute* over the connectives. The concept of distribution is the same as the one we find in daily speech, mathematics, and other theoretical subjects. For example, we know that multiplication distributes over addition. In first-order logic, the $\exists$-quantifier distributes over $\vee$-formulas, and the $\forall$-quantifier distributes over $\wedge$-formulas.

Q **Quantifiers and connectives.** Here are more examples of typical equivalences between first-order formulas:

$\exists x(Px \lor Qx)$ is equivalent to $\exists xPx \lor \exists xQx$.
If there is someone who dances or sings, then there is someone who dances or there is someone who sings, and vice versa.

$\forall x(Px \land Qx)$ is equivalent to $\forall xPx \land \forall xQx$.
If everyone dances and sings, then everyone dances and everyone sings, and vice versa.

$\forall x(Px \lor Qx)$ is *not* equivalent to $\forall xPx \lor \forall xQx$.
Everyone is a man or a woman, but not everyone is a man or everyone is a woman.

$\exists x(Px \land Qx)$ is *not* equivalent to $\exists xPx \land \exists xQx$.
There is an even number and an odd number, but there is no number that is both even and odd. ♦

> ## Digression
>
> The Greek philosopher *Aristotle* (384–322 B.C.E.) wrote a lot about logic and has had a tremendous influence on the development of science. In his work *De Interpretatione*, he discusses the relationship between quantified propositions and their negations. It can be summarized in the following **square of oppositions**:
>
>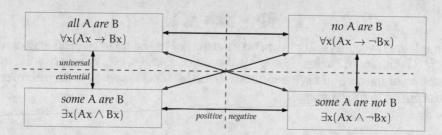
>
> We can read several properties out of this diagram: the propositions diagonally opposite each other are the negations of each other, the propositions in the top row cannot be true simultaneously, and the propositions in the bottom row cannot be false simultaneously. This relationship between propositions appears in many connections, for example between the concepts *valid*, *contradictory*, *satisfiable*, and *falsifiable*:
>
>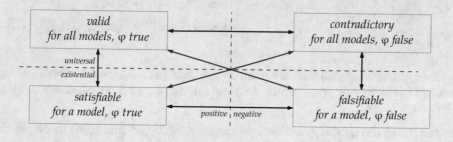

First-Order Logic and Modeling

First-order languages can be used to give precise descriptions of many types of structures. The action of using formulas to represent concrete structures or specific properties of structures is often referred to as *modeling*. Usually we have an intended model that we are trying to represent with formulas. In this process it is often necessary to check whether a formula is true in a given model. This is called **model checking** and is used extensively in computer science, for example when we want to check whether a program meets a given specification.

We are here going to look at a simple example of modeling by defining a language for talking about the appearance and location of some simple geometric figures. In this example, we assume that a, b, and c are constant symbols, and that *Circle*, *Square*, *Triangle*, *Big*, *Small*, and *SmallerThan* are relation symbols, all with arity one, except *SmallerThan*, which has arity two. We are not going to use any function symbols. We read atomic formulas in the following way:

$Circle(x)$	x is a circle	$Big(x)$	x is large
$Square(x)$	x is a square	$Small(x)$	x is small
$Triangle(x)$	x is a triangle	$SmallerThan(x, y)$	x is smaller than y

We are now going to make some different models for this language, and each has a subset of the following set as domain:

$$\{\bigcirc, \circ, \blacksquare, \square, \triangle, \vartriangle\}$$

The relation symbols will always be interpreted in the most natural way, that is, $Circle^{\mathcal{M}}$ is the set of circles, $Square^{\mathcal{M}}$ is the set of squares, etc. We specify the interpretation of the constant symbols for each model. This is sufficient for interpreting all formulas in the language.

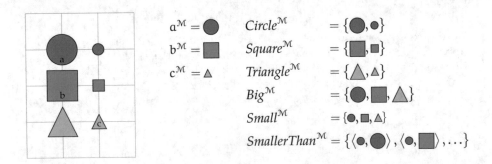

In this particular model, for example, *Circle*(a), *Square*(b), *Small*(c), and *SmallerThan*(c, b) are *true*, while *Triangle*(a), *Big*(c), and *SmallerThan*(a, b) are *false*.

We can make the language a little more interesting by adding some more relation symbols:

$Over(x, y)$	x *is closer to the top than* y
$Under(x, y)$	x *is closer to the bottom than* y
$LeftOf(x, y)$	x *is farther to the left than* y
$RightOf(x, y)$	x *is farther to the right than* y
$NextTo(x, y)$	x *is directly to the left of, to the right of, above, or below* y
$Between(x, y, z)$	x, y, *and* z *are in the same column, row, or diagonal, and* x *is between* y *and* z

We are now going to look at four different models by changing the interpretations of the constant and relation symbols:

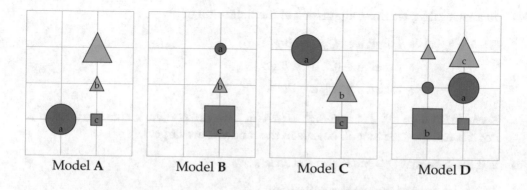

Model **A**   Model **B**   Model **C**   Model **D**

We can now reason about these models. If we interpret the relation symbols *Over* and *Under* in the model **A**, we get the following:

$$Over^{A} = \{ \, \langle \triangle, \triangle \rangle , \langle \triangle, \blacksquare \rangle , \langle \triangle, \bullet \rangle , \langle \triangle, \bullet \rangle , \langle \triangle, \blacksquare \rangle \, \}$$

$$Under^{A} = \{ \, \langle \triangle, \triangle \rangle , \langle \blacksquare, \triangle \rangle , \langle \bullet, \triangle \rangle , \langle \bullet, \triangle \rangle , \langle \blacksquare, \triangle \rangle \, \}$$

We see that the relations are the opposites of each other. In model **A**, we have that the formula $Over(b, c)$ is true. That is because $\langle b^{A}, c^{A} \rangle$, which is equal to $\langle \triangle, \blacksquare \rangle$, is in $Over^{A}$. We can therefore write $\mathbf{A} \models Over(b, c)$. Similarly, we get that $\mathbf{A} \models Under(a, b)$, because $\langle a^{A}, b^{A} \rangle$, which equals $\langle \bullet, \triangle \rangle$, is in $Under^{A}$.

Check for yourself that the following hold in the models **A** and **B**:

$\mathbf{A} \models Over(b, a)$	$\mathbf{B} \models \neg Over(b, a),$
$\mathbf{A} \models \neg Under(c, a),$	$\mathbf{B} \models Under(c, a),$
$\mathbf{A} \models LeftOf(a, b),$	$\mathbf{B} \models \neg LeftOf(a, b),$
$\mathbf{A} \models NextTo(a, c),$	$\mathbf{B} \models \neg NextTo(a, c),$
$\mathbf{A} \models \neg Between(b, a, c),$	$\mathbf{B} \models Between(b, a, c).$

❷   Is it the case that $\mathbf{B} \models \exists x Small(x)$?   ◆

❗ To answer that, we must look at the definition of ⊨:

$\mathbf{B} \models \exists x Small(x) \Leftrightarrow$ there is an $a \in |\mathbf{B}|$ such that $\mathbf{B} \models Small(\bar{a})$

$\qquad\qquad\qquad \Leftrightarrow$ there is an $a \in |\mathbf{B}|$ such that $\bar{a}^{\mathbf{B}} \in Small^{\mathbf{B}}$

$\qquad\qquad\qquad \Leftrightarrow$ there is an $a \in |\mathbf{B}|$ such that $a \in Small^{\mathbf{B}}$.

Because $Small^{\mathbf{B}} = \{●, ▲\}$, we can conclude that the answer is yes; there is a *small* object in the model $\mathbf{B}$.   ◆

❓ Is it the case that $\mathbf{C} \models \forall x Big(x)$?   ◆

❗ To answer that, we must again look at the definition of ⊨:

$\mathbf{C} \models \forall x Big(x) \Leftrightarrow$ for all $a \in |\mathbf{C}|$, $\mathbf{C} \models Big(\bar{a})$

$\qquad\qquad\qquad \Leftrightarrow$ for all $a \in |\mathbf{C}|$, $\bar{a}^{\mathbf{C}} \in Big^{\mathbf{C}}$

$\qquad\qquad\qquad \Leftrightarrow$ for all $a \in |\mathbf{C}|$, $a \in Big^{\mathbf{C}}$.

Because $|\mathbf{C}| = \{■, ⬤, ▲\}$ and $Big^{\mathbf{C}} = \{⬤, ▲\}$, we can conclude that the answer is no; it is not the case that all objects in the model $\mathbf{C}$ are large.   ◆

❓ Is it the case that $\mathbf{D} \models \forall x(Big(x) \rightarrow Circle(x))$?   ◆

❗ Again we need to look at the definition:

$\mathbf{D} \models \forall x(Big(x) \rightarrow Circle(x))$

$\qquad \Leftrightarrow$ for all $a \in |\mathbf{D}|$, $\mathbf{D} \models Big(\bar{a}) \rightarrow Circle(\bar{a})$

$\qquad \Leftrightarrow$ for all $a \in |\mathbf{D}|$, $\mathbf{D} \models Big(\bar{a})$ implies $\mathbf{D} \models Circle(\bar{a})$

$\qquad \Leftrightarrow$ for all $a \in |\mathbf{D}|$, if $a \in Big^{\mathbf{D}}$, then $a \in Circle^{\mathbf{D}}$

$\qquad \Leftrightarrow$ "all large objects are circles."

We can conclude that the answer is no; it is not the case that all large objects in the model $\mathbf{D}$ are circles. Both ▲ and ■ are counterexamples.   ◆

❓ Is it the case that $\mathbf{D} \models \forall x(Circle(x) \rightarrow \exists y \exists z Between(x, y, z))$?   ◆

❗ This exercise is more complex, but the procedure is the same:

$\mathbf{D} \models \forall x(Circle(x) \rightarrow \exists y \exists z Between(x, y, z))$

$\qquad \Leftrightarrow$ for all $a \in |\mathbf{D}|$, $\mathbf{D} \models Circle(\bar{a}) \rightarrow \exists y \exists z Between(\bar{a}, y, z)$

$\qquad \Leftrightarrow$ for all circles $a \in |\mathbf{D}|$, $\mathbf{D} \models \exists y \exists z Between(\bar{a}, y, z)$

$\qquad \Leftrightarrow$ for all circles $a \in |\mathbf{D}|$ then there are $b, c \in |\mathbf{D}|$ such that $\mathbf{D} \models Between(\bar{a}, \bar{b}, \bar{c})$

$\qquad \Leftrightarrow$ "all circles are between two objects in the model $\mathbf{D}$."

Yes, all circles are between two objects in the model $\mathbf{D}$. The only two circles are ● and ⬤ and we have that both $\langle ●, ■, ▲\rangle$ and $\langle ⬤, ■, ▲\rangle$ are elements of $Between^{\mathbf{D}}$.   ◆

❷ Are the following formulas satisfiable in the same model?

- *Big*(a) ∧ *Small*(b)
- ∀x*Triangle*(x)
- ∀x(*NextTo*(x, a) ∨ *NextTo*(x, b))
- ∀x(*Big*(x) → ∃y*Over*(y, x))
- ¬∃x(*LeftOf*(x, a) ∨ *RightOf*(x, a))   ◆

❶ Yes, here is a model:

❷ Is the formula ¬*Circle*(a) ∧ ¬*Triangle*(a) ∧ ¬*Square*(a) satisfiable?   ◆

❶ Yes, if we allow other types of objects in the models, the formula is satisfiable, for example, if |𝓜| = {⬠} and a^𝓜 = ⬠.   ◆

❷ Is the formula *Small*(a) ∧ *Big*(a) satisfiable?   ◆

❶ Yes, if we allow other types of objects in the models, this formula is satisfiable, for example, if |𝓜| = {🚲}, a^𝓜 = 🚲, and *Small*^{𝓜} = *Big*^{𝓜} = {🚲}.   ◆

❷ Provide a set of formulas that describes the following model as well as possible, that is, that has this model and essentially no others.

                                                                              ◆

Digression

The models of first-order logic interpret all symbols over one and the same set, but it is sometimes appropriate and useful to have different sets for different types of objects. In geometry, for example, it can be practical not to have points and lines in the same set. In **many-sorted logic** this is done by extending the signatures with **sorts**. This corresponds to *types* in typed programming languages, and such logics are sometimes called **typed** logics. In such systems all the terms must be defined with sorts, and in the semantics, these must be interpreted as elements of their respective sets. The models must therefore have one domain for each sort. It is possible to translate many-sorted logic to logic without sorts, but there are still many benefits to using sorts.

Theories and Axiomatizations

We can use what we have learned so far to clarify some well-known logical concepts.

Definition 16.3. Theories, axioms, and theorems

A **theory** is a set of formulas. The formulas of a theory are called **axioms**. All logical consequences of the theory are called **theorems**.

When a theory is given, we can look at the set of models that satisfies the theory. If these models have properties in common, we usually say that the theory **axiomatizes** these properties. It is also possible to do the opposite: when a model is given, we can look at the set of formulas that are true in the model. This is often called the **theory** of the model, and it can be used as a method for comparing models.

There is a limit to what it is possible to axiomatize with first-order logic. For example, it is not possible to give a complete characterization of all *finite* models by means of only first-order formulas. Such a characterization would – if it existed – have consisted of a set of formulas with the property that if a model made all the formulas true, then the model had to be finite, and conversely, that each model with a finite domain had to make all the formulas true. But such a set does not exist. It is possible to *prove* that such a set does not exist, and that it is therefore *impossible* to characterize all finite models in first-order logic, but we are not going to do this here. We say that finiteness is not *axiomatizable* in first-order logic.

Q   **The theory of equivalence relations.** Let T be the set consisting of the closed first-order formulas $\forall x Rxx$, $\forall x \forall y (Rxy \rightarrow Ryx)$, and $\forall x \forall y \forall z (Rxy \wedge Ryz \rightarrow Rxz)$. Each model that satisfies these formulas has to interpret R as an equivalence relation, and conversely, each model that interprets R as an equivalence relation has to make all the formulas true. We can express this in the following way:

$$\mathcal{M} \models T \Leftrightarrow R^{\mathcal{M}} \text{ is an equivalence relation.}$$

We say that the theory axiomatizes equivalence relations.   ◆

Some Technical Special Cases

Suppose that $\mathcal{M}$ is a model with domain D for a language in which f is a function symbol and R is a relation symbol. It is worthwhile to look at special cases that occur when the arity of these symbols is zero. It turns out that function symbols behave like constant symbols, and relation symbols like propositional variables:

– If the arity of f is zero, then by definition, we get that $f^{\mathcal{M}}$ is a function from D^0 to D, but what does that mean? Because D^0 consists of only one element $\langle \rangle$, the empty tuple, it follows that $f^{\mathcal{M}}$ also consists of only one element $\langle \langle \rangle , e \rangle$, where

Digression

There are many ways to axiomatize number theory. One such is called **Robinson arithmetic**, which is named after the American mathematician *Raphael Robinson* (1911–1995). The axioms are formulated in a first-order language defined by the signature $\langle 0 \,; s, +, \cdot \,; = \rangle$, where s, with arity one, represents the successor function, and + and $\cdot$, both with arity two, represent addition and multiplication:

(1)	$\forall x \neg (sx = 0)$	0 *is not the successor of any number*
(2)	$\forall x \forall y (sx = sy \rightarrow x = y)$	*a number has only one successor / s is injective*
(3)	$\forall x (x = 0 \vee \exists y (sy = x))$	*every number is either 0 or a successor*
(4)	$\forall x (x + 0 = x)$	x *plus zero is always* x
(5)	$\forall x \forall y (x + sy = s(x + y))$	*a recursive definition of +*
(6)	$\forall x (x \cdot 0 = 0)$	x *times zero is always zero*
(7)	$\forall x \forall y (x \cdot (sy) = xy + x)$	*a recursive definition of* $\cdot$

Robinson arithmetic is an interesting theory from a logical perspective, because it is just strong enough to have some very important properties. But the most famous axiomatization of number theory is perhaps **Peano arithmetic**, from 1889, named after the Italian mathematician *Giuseppe Peano* (1858–1932). We can define a theory that corresponds to Peano arithmetic from Robinson arithmetic by adding axioms that represent mathematical induction. For each formula Fx, we add

(8) $(F0 \wedge \forall x (Fx \rightarrow Fsx)) \rightarrow \forall x Fx$

Peano arithmetic thus has infinitely many axioms, and (8) is therefore called an **axiom schema**. This theory has been central in the development of mathematical logic and is still used as a way to define number theory.

$e \in D$. Therefore, we can identify $f^{\mathcal{M}}$ with e. There is no real difference between constant symbols and function symbols with arity zero, and we could, strictly speaking, do just fine without separate constant symbols. It is nevertheless practical not to have to refer to function symbols with arity zero.

- If the arity of R is zero, then by definition, we get that $R^{\mathcal{M}}$ is a subset of D^0. Because D^0 consists of only one element $\langle \rangle$, the empty tuple, there are exactly two possibilities for $R^{\mathcal{M}}$: it is either empty or equal to the set of the empty tuple, and in the first case, R is true in $\mathcal{M}$. So we see that R behaves exactly like a propositional variable in propositional logic.

Prenex Normal Form and More Equivalences

If x is a variable that does not occur freely in φ, then φ is equivalent to $\forall x \varphi$, because there is no effect in adding a quantifier. But we also have the following equivalences, which are useful if we want to move quantifiers:

$$\forall x(\varphi \wedge \psi) \quad \Leftrightarrow \quad \varphi \wedge \forall x\psi, \qquad\qquad \exists x(\varphi \wedge \psi) \quad \Leftrightarrow \quad \varphi \wedge \exists x\psi,$$
$$\forall x(\varphi \vee \psi) \quad \Leftrightarrow \quad \varphi \vee \forall x\psi, \qquad\qquad \exists x(\varphi \vee \psi) \quad \Leftrightarrow \quad \varphi \vee \exists x\psi,$$
$$\forall x(\varphi \rightarrow \psi) \quad \Leftrightarrow \quad \varphi \rightarrow \forall x\psi, \qquad\qquad \exists x(\varphi \rightarrow \psi) \quad \Leftrightarrow \quad \varphi \rightarrow \exists x\psi,$$
$$\forall x(\psi \rightarrow \varphi) \quad \Leftrightarrow \quad \exists x\psi \rightarrow \varphi, \qquad\qquad \exists x(\psi \rightarrow \varphi) \quad \Leftrightarrow \quad \forall x\psi \rightarrow \varphi.$$

In several situations it is useful to transform formulas into a specific *standard* form before we do anything else with them. If we are creating a computer program that does something with formulas, this is often the first step. One such form is called *prenex normal form*.

Definition 16.4. Prenex normal form

A closed first-order formula is in **prenex normal form** if it is of the form $Q_1x_1Q_2x_2 \ldots Q_nx_n\varphi$, where each Q_i is either $\forall$ or $\exists$, and φ is without quantifiers.

In prenex normal form	*Not in prenex normal form*
$Pa \wedge Pb$	$Pa \wedge \forall xQx$
$\forall xPx$	$\forall x(Px \rightarrow \exists yRxy)$
$\forall x\exists y(Rxy \wedge Ryx)$	$\forall x(\forall yRxy \rightarrow \forall zSxz)$

It is the case that every closed first-order formula is equivalent to a formula in prenex normal form. If φ is a closed first-order formula, we can construct an equivalent formula in prenex normal form in the following way: First, rename all variables such that no quantifiers bind the same variable. Then move the quantifiers outward using the above equivalences.

Q **Prenex normal form.** The formula $\forall xPx \vee \exists yQy$ is not in prenex normal form, but we can use two of the above equivalences to put it in prenex normal form:

$$\forall xPx \vee \exists yQy \quad \Leftrightarrow \quad \forall x(Px \vee \exists yQy) \quad \Leftrightarrow \quad \forall x\exists y(Px \vee Qy). \qquad \blacklozenge$$

Final Comments

By means of first-order languages we have obtained a logic with great expressivity and the possibility of representing complex connections, because models may consist of functions and relations in a rich and complex interplay. We have only just begun to explore the relationship between first-order languages and models, but we must stop here. We could have gone further with questions such as, "when do theories have isomorphic models?" and "what is it that can and cannot be axiomatized in first-order logic?" And we could have looked at the strength and the properties of concrete theories. We shall return to logic in Chapter 24, where we look at *logical calculi*.

Exercises

16.1 Why are all valid formulas equivalent to each other?

16.2 Find first-order formulas that represent the following set-theoretic statements and decide whether each is valid:

(a) $A \subseteq B$
(b) $a \in (A \cap (B \cup C))$
(c) $(A \cup B) \subseteq (A \cap B)$
(d) $A \cup (B \cap C) \subseteq (A \cup B) \cap (A \cup C)$

16.3 Why is $\forall x \exists y Rxy$ not the same as $\exists y \forall x Rxy$?

16.4 Suppose that the following drawing represents the model $\mathcal{M}$ and that all symbols are interpreted in the natural way.

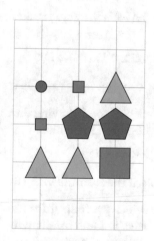

Find the truth values of the following formulas:

(a) $\exists x(\textit{Small}(x) \wedge \textit{Triangle}(x))$
(b) $\exists x(\textit{Small}(x) \wedge \textit{Square}(x))$
(c) $\forall x(\textit{Small}(x) \rightarrow \textit{Square}(x))$
(d) $\forall x(\textit{Circle}(x) \rightarrow \textit{Small}(x))$
(e) $\forall x(\textit{Triangle}(x) \rightarrow \textit{Big}(x))$
(f) $\exists x \neg \exists y(\textit{Under}(x,y))$
(g) $\exists y \forall x(\textit{Over}(x,y))$
(h) $\forall x(\textit{Small}(x) \rightarrow \exists y(\textit{Big}(y) \wedge \textit{Over}(x,y)))$
(i) $\exists x \exists y(\textit{Square}(x) \wedge \textit{Big}(y) \wedge \textit{Under}(x,y))$

16.5 Prove the following statements:

(a) $\neg \forall x F$ is *not* equivalent to $\forall x \neg F$
(b) $\neg \exists x F$ is *not* equivalent to $\exists x \neg F$
(c) $\forall x(Px \vee Qx)$ is *not* equivalent to $\forall x Px \vee \forall x Qx$
(d) $\exists x(Px \wedge Qx)$ is *not* equivalent to $\exists x Px \wedge \exists x Qx$
(e) $\exists x(Px \rightarrow Qx)$ is *not* equivalent to $\exists x(\neg Px \wedge Qx)$

16.6 Decide which of the following formulas are logical consequences of which:

$$\forall x Px \qquad\qquad \exists x Px \qquad\qquad Pa \wedge Pb$$

16.7 Decide which of the following formulas are logical consequences of which:

$$\forall x \exists y Rxy \qquad \exists y \forall x Rxy \qquad \forall x \forall y Rxy \qquad \exists x \exists y Rxy$$

16.8 Decide which of the following formulas are logical consequences of which:

$$\exists x(Px \wedge Qx) \qquad \exists xPx \wedge \exists xQx \qquad \exists xPx \qquad \exists xPx \vee \exists xQx$$

16.9 Decide which of the following formulas are logical consequences of which:

$$\forall x(Px \vee Qx) \qquad \forall xPx \vee \forall xQx \qquad \forall xPx \qquad \forall xPx \wedge \forall xQx$$

16.10 Prove that $(A \rightarrow B) \wedge (B \rightarrow A)$ is valid if and only if the formulas A and B are equivalent.

16.11 Prove that the formula $\forall x\forall y\forall z(Rxy \wedge Ryz \rightarrow Rxz)$ is true in a model $\mathcal{M}$ *if and only if* R is interpreted as a transitive relation.

16.12 Suppose that it is not the case that there is not someone who has the property P. Does that mean that everyone has the property P?

16.13 Are the following two formulas equivalent? Provide a proof or find a counterexample.

(a) $\forall xPx \wedge \forall x(Px \rightarrow Qx)$ \qquad\qquad (b) $\forall xQx$

16.14 Suppose that $\mathcal{M} \models \forall xPx$. Is it then the case that $\mathcal{M} \models \forall x(Px \vee Qx)$?

16.15 For each of the following formulas, find an equivalent formula that does *not* contain any negation symbols:

(a) $\neg\forall x\forall y\neg Rxy$ \qquad (b) $\neg\exists x\forall y\neg Rxy$ \qquad (c) $\neg\exists x(Px \wedge \neg Qx)$

16.16 Look at the formula $\forall x\exists y(Rxy \wedge \neg Ryx)$.

(a) Specify a first-order model $\mathcal{M}$ with domain $\{1, 2, 3\}$ that makes this formula *true*. Explain briefly why the model makes the formula true by referring to the definition of the interpretation of first-order formulas.

(b) Can R in the previous problem be interpreted as a reflexive relation? If yes, explain briefly how this is possible; if no, explain why it is not possible.

(c) Here are some first-order formulas. Place $\Rightarrow$-arrows that indicate which formulas are logical consequences of which. For example, place an arrow from F to G if G is a logical consequence of F. It is not necessary to place an arrow from a formula to itself.

$$\forall x(Px \vee Qx) \qquad\qquad \forall xPx \vee \forall xQx$$

$$\forall xPx \wedge \forall xQx \qquad\qquad \forall xPx$$

16.17 Show that $\exists x(Px \rightarrow Qx)$ is equivalent to $\forall xPx \rightarrow \exists xQx$.

16.18 Transform $Pa \wedge (\forall xQx \vee \forall xRx)$ to prenex normal form.

Chapter 17

Abstraction with Equivalences and Partitions

■■▪■■▪■■■▪■■▪■■▪■▪■▪■■▪■■

In this chapter you will learn about how equivalence relations, equivalence classes, and partitions are related to each other, and how these can be used for abstraction.

Abstracting with Equivalence Relations

In mathematics we observe that

$$2+2 = 2 \cdot 2, \qquad 3+3 = 2 \cdot 3, \qquad 4+4 = 2 \cdot 4, \qquad 5+5 = 2 \cdot 5,$$

and we *abstract* over these concrete numbers and conclude that $x + x = 2 \cdot x$. However, we abstract all the time, and not only in mathematics. For example, there are innumerable ways to write the letter s:

$$s, \mathbf{s}, s, \mathbb{S}, \mathbf{S}, \mathfrak{S}, \mathcal{S}, \mathsf{S}, \mathscr{S}, \mathsf{S}, s, \dots$$

We have *abstracted* over the shape of all the characters we have seen, and when we come across new characters, we recognize them, even though we have never seen them before. We are even able to distinguish the letter S from the number 5. In mathematical terms, we have put an equivalence relation on characters; we perceive some characters as equivalent, and we make no distinction between them. When looking for the essence of something, we disregard and abstract out details that are of no interest to us. In the next example, you will find further illustrations of abstractions that we make in daily life.

Q **Abstractions and equivalence relations.** A bouncer tends to focus on whether the guests are above or below a certain age, and whether they are drunk or not, but considers details such as hair color and education to be irrelevant. The bouncer abstracts away those details and considers those who are *equally old* to be equivalent to each other, and likewise for those who are *equally drunk*. When we refer to *action movies* or *romantic comedies*, we abstract out the details that concern actors, language, the year, and other characteristics. When we talk about *coffee*, we abstract over all variants of coffee. When we say, "these computer programs all do the same thing" or "this person is just as cute as that one," we identify computer programs or people on the basis of the properties that interest us in a given context. ♦

© Springer Nature Switzerland AG 2021
R. Antonsen, *Logical Methods*, https://doi.org/10.1007/978-3-030-63777-4_18

An equivalence relation, as introduced in Chapter 6 (*page 70*), is a form of identification between elements and a way to make abstraction explicit. It tells us which elements are equivalent and which are not. Using an equivalence relation is similar to wearing colored glasses; you can no longer tell certain colors apart.

Equivalence Classes

In practice, we use equivalence relations to identify elements that we do not want to tell apart. Collecting all equivalent elements of a single set is so useful that it has been given its own name: an *equivalence class*. The advantage of equivalence classes is that they permit us to handle everything in the same equivalence class in an identical – and *equivalent* – manner.

Definition 17.1. Equivalence class

Assume that $\sim$ is an equivalence relation on a set S. We say that the **equivalence class** of an element $x \in S$ is the set $\{y \in S \mid y \sim x\}$, in other words, the set of elements of S that are related to x. We write $[x]$ for the equivalence class of x, and $S/\sim$ for the set of all equivalence classes. This set is called the **quotient set** of S under $\sim$.

Q **Equivalence class.** Let $\sim$ be the relation $\{\langle 1,1 \rangle, \langle 2,1 \rangle, \langle 1,2 \rangle, \langle 2,2 \rangle, \langle 3,3 \rangle\}$ on the set $S = \{1, 2, 3\}$. We can draw it like this: 👓 👓. This is an equivalence relation on S, because it is reflexive, symmetric, and transitive. The definition tells us that the equivalence class of 1, denoted by $[1]$, is equal to $\{1, 2\}$, the equivalence class of 2, denoted by $[2]$, is equal to $\{1, 2\}$, and the equivalence class of 3, denoted by $[3]$, is equal to $\{3\}$. The set of all equivalence classes, the quotient set of S, is equal to $S/\sim = \{[1], [2], [3]\} = \{\{1, 2\}, \{3\}\}$.   ◆

Q **The same age as.** Assume that S is the set of all people currently alive, and that $x \sim y$ means that x *is the same age as* y. This is an equivalence relation on the set of all people currently alive, because it is reflexive, symmetric, and transitive. Here we have exactly one equivalence class for each age. For example, all those who are 18 years old are equivalent to each other, and only to each other, and they end up in the same equivalence class.   ◆

The elements that are equivalent to each other, and that we for the occasion are identifying, are all in the same equivalence class. Proceeding from a set M via an equivalence relation on M to the set of corresponding equivalence classes is a powerful mathematical tool. If we have a set that for some reason is too large or too fine-grained, we can replace it with the set of equivalence classes for a suitable equivalence relation. There are two extremes here. The first extreme is the identity relation, which relates elements only to themselves. As a result, we have one equivalence class for each element, or in other words, no essential change. The second extreme is the universal relation, which relates all of the elements to each other. As a result, we have a single equivalence class that contains all of the elements, which will be totally useless in most cases.

Q Binary numbers as equivalence classes. A natural equivalence relation on the set of *bit strings* is the one saying that two bit strings, for example 01 and 001, are equivalent if they represent the same numeric value. The set of equivalence classes of bit strings can be regarded as the set of **binary numbers**.   ◆

Q Integers modulo n. Let n be a positive integer. We define an equivalence relation ≡ on the set of integers by saying that $x \equiv y$ holds when x and y have the same remainder when we divide by n, or in other words, that $x - y$ is divisible by n. This can be written $x \equiv y \pmod n$, and we say that "x is congruent to y modulo n." For example, if $n = 5$, we say that $7 \equiv 12 \pmod 5$, because both 7 and 12 give the remainder 2 on division by 5, and that $8 \equiv 13 \pmod 5$ because both 8 and 13 give the remainder 3 when we divide by 5. This gives rise to the following equivalence classes:

$$[0] = \{\ldots, -5, 0, 5, 10, 15, \ldots\},$$
$$[1] = \{\ldots, -4, 1, 6, 11, 16, \ldots\},$$
$$[2] = \{\ldots, -3, 2, 7, 12, 17, \ldots\},$$
$$[3] = \{\ldots, -2, 3, 8, 13, 18, \ldots\},$$
$$[4] = \{\ldots, -1, 4, 9, 14, 19, \ldots\}.$$

Such sets are called **residue classes** and **congruence classes**, and the set of them, $\mathbb{Z}/\equiv$, is called the set of **integers modulo** n and denoted by $\mathbb{Z}/n$ or $\mathbb{Z}/n\mathbb{Z}$.   ◆

Q Circular strings. Assume that we want to identify strings that become equal if we *rotate* the positions of the characters. We want, for example, to identify the strings 1234, 2341, 3412, and 4123 because they are rotations of each other, but not 1234 with 4321, because they can never become equal by rotation. *How can we do this?* One answer is to define *equivalence classes* of strings: define a relation ∩ for strings so that (s ∩ t) holds exactly when t is obtained from s by moving the leftmost character to the rightmost position. We can then see that (xs ∩ sx) holds for all strings s and characters x. For example, we see that 1234 ∩ 2341 and 1000 ∩ 0001. However, this relation is generally *not* an equivalence relation. Therefore, let ↺ be the transitive closure of ∩. Check for yourself that this relation is now an equivalence relation. Now we get, for example, the equivalence classes [12] = {12, 21} and [123] = {123, 231, 312}, enabling us to write the identities [123] = [231] and [123] = [312]. It is worth rethinking what happens here: we started from an alphabet A, but the set A^* of all strings over A contained more than what we were looking for. Therefore, we defined an equivalence relation and constructed a set that suited our needs. The new set is the set of all equivalence classes $A^*/↺$, called the quotient set of A^* under ↺.   ◆

Digression

There are mathematical notation systems for describing and calculating most things. A nice example of this is *notation systems for juggling patterns*. What characterizes a good notation system is that it abstracts just the right amount. For example, when it comes to juggling, it is not so interesting exactly how high the objects, let us say balls, are thrown or where the hands are in relation to each other. A notation system called "siteswap" captures the *order* of juggling throws and identifies a juggling pattern with a number sequence. The number sequence 333..., which consists only of 3's, represents ordinary juggling with three balls. Each digit indicates how many beats it takes before a ball is rethrown, under the assumption that the hands throw alternately and that the balls are thrown in an even tempo. We can make a diagram that shows the paths of the balls in this way:

Here are 441, 450, 51414, and 531, which are juggling patterns for three balls:

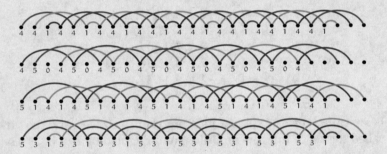

And here is one called 534 for four balls:

A lot of what we have learned can be used for *juggling mathematics*: For example, we can define an *equivalence relation* on juggling patterns such that 534, 345, 453, and 534534 become equivalent, as they should be, because they are rotations or repetitions of each other. In the general case, a *juggling pattern* is defined as a *bijection* on the integers, something that is apparent in the above diagrams. It is also possible to prove statements like *the average of the digits in a juggling pattern equals the number of objects juggled*, and much more.

Partitions

Often when we think of sets, we divide the elements into natural subsets, in which distinguishing between the elements of each subset is of less interest. For example, the set of university students is divided into bachelor's and master's students. The set of integers is divided into even and odd numbers. The set of propositional formulas is divided into valid and falsifiable formulas. The characteristic of all these subdivisions is that they encompass the entire underlying set without overlapping. This is made completely precise in the concept of a *partition*.

Definition 17.2. Partition

A **partition** of a set S is a set X of nonempty subsets of S that satisfies the following conditions:

- The union of all the sets of X equals S.
- The intersection of two different sets from X is empty.

In other words: If $S_1, S_2, \ldots, S_n$ are nonempty subsets of S, S equals $S_1 \cup S_2 \cup \cdots \cup S_n$, and for all S_i and S_j such that $i \neq j$, it is true that $S_i \cap S_j = \emptyset$, then $\{S_1, S_2, \ldots, S_n\}$ is a partition of S. The first condition is that "the sets cover all of S." The second condition is that the sets are pairwise **disjoint** or mutually exclusive.

Q **Partitions.** Let $S = \{a, b, c\}$. There are exactly five partitions of S, which are $\{\{a, b, c\}\}$, $\{\{a\}, \{b, c\}\}$, $\{\{b\}, \{a, c\}\}$, $\{\{c\}, \{a, b\}\}$,   and $\{\{a\}, \{b\}, \{c\}\}$.   ◆

The term "partition" may appear confusing at first, but a partition is the *set* of subsets that satisfies the conditions above, and not the subsets of S themselves. The term is derived from the Latin noun *partitio*, which means *splitting up, division*, or *classification*. When we talk about different partitions, we refer to the various possible ways of dividing up a set.

Q **Partition, number sets.** Let $\mathbb{Z}$ be the set of integers, let P be the set of even numbers, and let O be the set of odd numbers. Since $P \cup O = \mathbb{Z}$ and $P \cap O = \emptyset$, $\{P, O\}$ is a partition of $\mathbb{Z}$.   ◆

Q **Partition, propositional formulas.** Let U be the set of propositional formulas, let G be the set of valid propositional formulas, and let F be the set of falsifiable propositional formulas. Because all formulas are either valid or falsifiable, $G \cup F = U$, and no formulas are both valid and falsifiable, $G \cap F = \emptyset$, the set that comprises G and F, namely the set $\{G, F\}$, is a partition of U.   ◆

❷ How many partitions does a set of four elements have?   ◆

A set of four elements has fifteen partitions. The fifteen partitions of $\{1, 2, 3, 4\}$ are the following:

$$
\begin{array}{lllll}
\{\{1\},\{2\},\{3\},\{4\}\}, & \{\{1\},\{4\},\{2,3\}\}, & \{\{1,3\},\{2\},\{4\}\}, & \{\{1,3\},\{2,4\}\}, & \{\{1,2,4\},\{3\}\}, \\
\{\{1\},\{2\},\{3,4\}\}, & \{\{1\},\{2,3,4\}\}, & \{\{1,4\},\{2\},\{3\}\}, & \{\{1,4\},\{2,3\}\}, & \{\{1,3,4\},\{2\}\}, \\
\{\{1\},\{3\},\{2,4\}\}, & \{\{1,2\},\{3\},\{4\}\}, & \{\{1,2\},\{3,4\}\}, & \{\{1,2,3\},\{4\}\} & \{\{1,2,3,4\}\}.
\end{array}
$$

A natural question is whether there is any structure on the set of all partitions. We shall now see that we can *order* the set by calling one partition *finer* than another.

Definition 17.3. Refinement of a partition

Let X and Y be partitions of a set M. If each element of X is a subset of an element of Y, we write $X \leqslant Y$ and say that X is a **refinement** of Y and is **finer** than Y.

Q   Refinement of partition. The partition $\{\{1\}, \{2\}, \{3, 4\}\}$ is finer than the partition $\{\{1, 2\}, \{3, 4\}\}$. This is because all the elements of the set $\{\{1\}, \{2\}, \{3, 4\}\}$ are subsets of elements of the set $\{\{1, 2\}, \{3, 4\}\}$. Both $\{1\}$ and $\{2\}$ are subsets of $\{1, 2\}$, and $\{3, 4\}$ is a subset of $\{3, 4\}$.

This relation is an example of a *partial order*, which means that it is reflexive, transitive, and antisymmetric. Here follows an exercise about this with a complete solution. You should try to solve the problem on your own before reading the solution.

? (a) What are the finest and coarsest (least fine) partitions of $\{a, b, c, d, e\}$?

   (b) Show that $\leqslant$ is reflexive.

   (c) Show that $\leqslant$ is transitive.

   (d) Show that $\leqslant$ is antisymmetric.

! (a) The finest partition of $\{a, b, c, d, e\}$ is $\{\{a\}, \{b\}, \{c\}, \{d\}, \{e\}\}$. The coarsest partition of $\{a, b, c, d, e\}$ is $\{\{a, b, c, d, e\}\}$.

   (b) In order to show that $\leqslant$ is reflexive, suppose that X is a partition of a set M. Because each element of X is a subset of itself, we have that $X \leqslant X$.

   (c) In order to show that $\leqslant$ is transitive, suppose that $X \leqslant Y$ and $Y \leqslant Z$. We must show from this assumption that $X \leqslant Z$. In order to show that $X \leqslant Z$, suppose that $x \in X$. We must show that there is a set $z \in Z$ such that $x \subseteq z$. From $x \in X$ and the assumption that $X \leqslant Y$, it follows that there is some $y \in Y$ such that $x \subseteq y$. From $y \in Y$ and the assumption that $Y \leqslant Z$, it follows that there is some $z \in Z$ such that $y \subseteq z$. Because $x \subseteq y$ and $y \subseteq z$, it follows that $x \subseteq z$, which was what we wanted to show.

(d) In order to show that $\leqslant$ is antisymmetric, suppose that $X \leqslant Y$ and $Y \leqslant X$. We must show from this assumption that $X = Y$. Therefore, assume that $x \in X$. It suffices to show that $x \in Y$, because then we will have shown that $X \subseteq Y$, and the proof for $Y \subseteq X$ is done similarly. Because $x \in X$ and $X \leqslant Y$, there is an element $y \in Y$ such that $x \subseteq y$. Because $Y \leqslant X$, there is also an element $x' \in X$ such that $y \subseteq x'$. By transitivity of $\subseteq$ it follows that $x \subseteq x'$. Because X is a partition, x and x' cannot be different sets, because in that case, X would contain overlapping sets. It follows from $x \subseteq y \subseteq x$ that $x = y$, and hence that $x \in Y$. ◆

Q **Partial order of partitions.** We can draw all the partitions of $\{1, 2, 3, 4\}$ in the following way, where a line from X up to Y means that $X \leqslant Y$. This is yet another example of a Hasse diagram, like the one we saw in Chapter 6 (*page 73*).

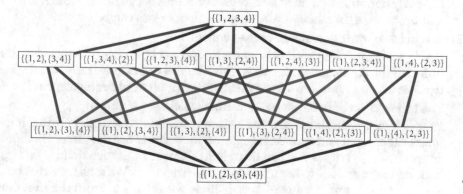

Chapter 6 (*page 73*)

Digression

Take a list of positive integers: $1, 2, 4, 1, 4, 8$. *Is it possible to separate it into two lists such that the sum of the numbers in each list is the same?* This is called the **partition problem** and is one of the famous **NP**-complete problems. In this case, there are many possibilities, for example $1, 1, 4, 4$ and $2, 8$, which both sum to 10, but in the general case, no one has found an efficient method that works for every list.

The Connection Between Equivalence Classes and Partitions

There is a beautiful connection between partitions and equivalence classes: *If we take the set of all equivalence classes for a given equivalence relation, we get a partition.* An equivalence relation thus divides a set into equivalence classes that make up a partition. We shall prove this now, and the first step of the proof is to show that if two elements are related to each other, then they are in the same equivalence class. Try to solve the next exercise yourself before continuing to read.

❷ Let $\sim$ be an equivalence relation on a set S, and suppose that $x \sim y$. Show that $[x] = [y]$. ◆

● We suppose, as given, that ~ is an equivalence relation on S, and that $x \sim y$. In order to show that $[x] = [y]$, it is sufficient to show that $[x] \subseteq [y]$ and $[y] \subseteq [x]$. To show that $[x] \subseteq [y]$, suppose that $z \in [x]$, for an arbitrarily chosen element z. We must show from this assumption that $z \in [y]$. The definition of $z \in [x]$ gives that $z \sim x$. We then have both $z \sim x$ and $x \sim y$, and because ~ is transitive, $z \sim y$. The definition of $[y]$ gives that $z \in [y]$. We have then shown that $[x] \subseteq [y]$. The proof of $[y] \subseteq [x]$ is similar, where we have to use that ~ is a symmetric relation.   ◆

We shall now prove that if ~ is an equivalence relation on a set S, then the set of all equivalence classes is a partition of S. The proof consists of three parts, one for each part of the definition of a partition:

(1) Each equivalence class $[x]$ is a nonempty subset of S.
    Proof: By definition, $[x]$ must be a subset of S, and because ~ is reflexive, $x \sim x$, and then $x \in [x]$; in other words, no equivalence class can be empty.

(2) The union of the equivalence classes equals S.
    Proof: Each equivalence class is by definition a subset of S, and therefore the union of the equivalence classes is also a subset of S. Conversely, $x \in S$ gives that $x \in [x]$, and therefore S is a subset of the union of the equivalence classes.

(3) The intersection of two different equivalence classes is empty.
    Proof: Here, it is easier to show the contrapositive statement, namely that if the intersection is not empty, then the equivalence classes are the same. Therefore, suppose that $[x] \cap [y] \neq \varnothing$. Then there is an element $z \in S$ such that $z \in [x] \cap [y]$, and then $z \sim x$ and $z \sim y$. Because ~ is symmetric, $x \sim z$. We thus have that both $x \sim z$ and $z \sim y$, and because ~ is transitive, we get $x \sim y$. From the previous exercise we know that the equivalence classes must be identical, $[x] = [y]$.

We have thus proved that the set of equivalence classes constitutes a partition. But the converse is also true; if we have a partition, we can define an equivalence relation such that two elements are related to each other precisely when they are both in the same subset of the partition. This relation is an equivalence relation, and the set of equivalence classes equals the partition we began with.

Q **Equivalence classes.** Let $S = \{1, 2, 3, 4, 5, 6\}$, let $A = \{1, 3, 5\}$, and let $B = \{2, 4, 6\}$. A partition of S is $\{\{1, 3, 5\}, \{2, 4, 6\}\}$ or $\{A, B\}$. We can now define the relation R by saying that $\langle x, y \rangle \in R$ if x and y are in the same subset of the partition, either A or B. For example, $\langle 1, 3 \rangle \in R$, because both 1 and 3 are in A, but $\langle 1, 2 \rangle \notin R$, because 1 is in A and 2 is in B. The definition of R immediately gives that A and B are the equivalence classes. We see that the set of equivalence classes $\{A, B\}$ equals the partition that we started with.   ◆

Digression

Japan has a long tradition of incense called kōdō (香道), and one of the first known practical applications of the concept of a *partition* comes from a game called Genji-kō (源氏香). The game is played by taking five different types of incense and making five small packets of each type, so that there are 25 packets in all. Then five packets are selected at random and smelled. The task is to determine which of the scents are the same. In order to represent the answer, a tradition emerged whereby five vertical lines were drawn for each scent, as well as horizontal lines between the tops, to indicate which scents were the same. For example, ▥ indicates that the first two and the last three scents are the same. Here are all 52 possibilities, which correspond to the partitions of the set $\{1, 2, 3, 4, 5\}$. At the top left is the partition $\{\{1\}, \{2\}, \{3\}, \{4\}, \{5\}\}$, representing five different scents, and at the bottom right is the partition $\{\{1, 2, 3, 4, 5\}\}$, representing that all the scents are the same.

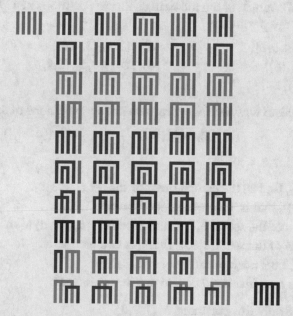

These symbols are called Genji-mon (源氏 紋), because they were used as symbols for the different chapters of *The Tale of Genji* (源氏物 語), a classic work of Japanese literature from around the year 1021. However, the story consists of 54 chapters, and therefore two of the partitions were used twice, but drawn differently. Here is a similar visualization of the fifteen different partitions of $\{1, 2, 3, 4\}$:

Exercises

17.1 In this exercise you will find examples of partitions.

(a) Find a partition of the set $\{1, 2, a, b\}$.

(b) Find a different partition of the set $\{1, 2, a, b\}$.

(c) Find a partition of the set $\{a, b, c, d, e, f, g, h\}$ with $\{a, b, c, d\}$ as an element.

(d) Find a partition of the natural numbers with four elements.

(e) Find a partition of the natural numbers with infinitely many elements.

(f) Find a partition of propositional formulas with four elements.

17.2 Let X be the set of all the people you know, and find at least three natural partitions of this set. Provide reasons for why these are partitions; use the definition of partition in your justification.

17.3 Let $S = \{1, 2, 3, 4, 5, 6\}$. Which of the following sets are partitions of S? Explain why or why not the sets are partitions:

(a) $\{\{1, 2\}, \{2, 3, 4\}, \{4, 5, 6\}\}$

(b) $\{\{1\}, \{2, 5, 3\}, \{4\}, \{6\}\}$

(c) $\{\{1, 3, 5\}, \{2, 4, 6\}\}$

(d) $\{\{1, 2\}, \{3, 4, 6\}\}$

(e) $\{\varnothing, \{1, 3, 5\}, \{2, 4, 6\}\}$

17.4 Let $S = \{a, b, c\}$. Explain why the following sets are *not* partitions of S:

(a) $\{\{a\}, \{c\}\}$

(b) $\{\varnothing, \{a, b\}, \{c\}\}$

(c) $\{\{a, b\}, \{b, c\}\}$

17.5 Suppose that $U = \{1, 2, 3, 4, 5\}$.

(a) What is $\overline{\{1, 2, 3\}}$, that is, the complement of the set $\{1, 2, 3\}$?

(b) What is $\mathcal{P}(\{1, 4\})$, that is, the power set of the set $\{1, 4\}$?

(c) Find a partition of the set $\{a, b, c, d, e, f\}$ in which $\{a, b, c, d\}$ is an element.

(d) Find a partition of the set $\{1, 2, 3, 4\}$ that has two elements.

(e) Why is $\{\{1, 2\}, \{2, 3\}\}$ not a partition of $\{1, 2, 3\}$?

(f) Is it always the case that $X \in \mathcal{P}(X)$, no matter what X is?

17.6 Show that there are only five partitions of $\{1, 2, 3\}$.

17.7 Is there a partition of the empty set? If so, what is it? Hint: Read the definition of partition carefully.

17.8 Let R be a transitive and symmetric relation on the set S. Suppose that for all $x \in S$, there exists $y \in S$ such that xRy. Prove that R is an equivalence relation.

17.9 Let $x \sim y$ mean that x *is the same age as* y. Show that this is an equivalence relation. What are the equivalence classes?

17.10 Suppose that an equivalence relation R is given, and that a is an element of the equivalence class of b. Explain why it must be the case that aRb.

17.11 Let X be the set of all the movies you have ever seen. Define three different equivalence relations on X, and describe the equivalence classes that these relations give rise to.

17.12 Let $S = \{a, b, c, d, e, f, g\}$.

(a) Find a set P such that P is a partition of this set.

(b) Explain why P is a partition.

(c) Assume that there is an equivalence relation such that each element of P is an equivalence class. What is this equivalence relation?

(d) What is the equivalence class of a?

(e) What is the equivalence class of b?

17.13 Decide whether the following can be partitions. If so, what is the equivalence relation that gives rise to the partition?

(a) $\{\{1\}, \{2\}, \{3\}\}$ (b) $\{\{1\}, \{2, 3\}\}$ (c) $\{\{1, 2\}, \{2, 3\}\}$

17.14 Below, some equivalence relations on $\{1, 2, 3, 4\}$ are listed. For each of them, find $[1]$, that is, the equivalence class of 1:

(a) $\{\langle 1, 1 \rangle, \langle 2, 2 \rangle, \langle 3, 3 \rangle, \langle 4, 4 \rangle\}$

(b) $\{\langle 1, 1 \rangle, \langle 2, 2 \rangle, \langle 3, 3 \rangle, \langle 4, 4 \rangle, \langle 1, 2 \rangle, \langle 2, 1 \rangle\}$

(c) $\{\langle 1, 1 \rangle, \langle 2, 2 \rangle, \langle 3, 3 \rangle, \langle 4, 4 \rangle, \langle 1, 3 \rangle, \langle 3, 1 \rangle\}$

(d) $\{\langle 1, 1 \rangle, \langle 2, 2 \rangle, \langle 3, 3 \rangle, \langle 4, 4 \rangle, \langle 1, 4 \rangle, \langle 4, 1 \rangle, \langle 2, 3 \rangle, \langle 3, 2 \rangle\}$

(e) $\{\langle 1, 1 \rangle, \langle 2, 2 \rangle, \langle 3, 3 \rangle, \langle 4, 4 \rangle, \langle 2, 3 \rangle, \langle 3, 2 \rangle\}$

(f) $\{\langle 1, 1 \rangle, \langle 2, 2 \rangle, \langle 3, 3 \rangle, \langle 4, 4 \rangle, \langle 1, 2 \rangle, \langle 1, 4 \rangle, \langle 2, 1 \rangle, \langle 2, 4 \rangle, \langle 4, 1 \rangle, \langle 4, 2 \rangle\}$

17.15 Suppose that $[x] = [y]$ for a given equivalence relation. What can you say about x and y? For example, is it the case that $x = y$?

17.16 We have previously shown that $\Leftrightarrow$ is an equivalence relation on the set of propositional formulas.

(a) Describe briefly the equivalence classes for this relation.

(b) What are the equivalence classes of $\top$ and $\bot$?

(c) Is there an equivalence class with only one element?

(d) Is there an equivalence class in which both P and Q are elements?

(e) Are the formulas $P \lor Q$ and $P \land Q$ in different equivalence classes?

17.17 Describe the equivalence classes for the following equivalence relations.

(a) The equality relation on the natural numbers, that is, the relation $=$.

(b) The relation $\sim$ on the integers such that $x \sim y$ if $x + y$ is an even number.

(c) The "has the same mother as" relation on the set of all people.

(d) The "has the same number of letters as" relation on the set of words.

17.18   Let the relation $\sim$ be defined such that $S \sim T$ if $|S| = |T|$. Show that $\sim$ is an equivalence relation. What are the equivalence classes?

17.19   Let $\sim$ be a relation on the integers such that $x \sim y$ if x and y have the same remainder when divided by 4. For example, $5 \sim 9$ because both 5 and 9 have remainder 1 when divided by 4. Describe the equivalence classes of this relation.

17.20   Explain in your own words why two elements that are not equivalent end up in different equivalence classes. Feel free to use an example.

17.21   Show that all the elements of an equivalence class are related to each other. In other words: Let $\sim$ be an equivalence relation on a set S, and let $[a]$ be the equivalence class of $a \in S$. Show that if $x, y \in [a]$, then $x \sim y$.

17.22   Let $\sim$ be an equivalence relation on a set S, and let $[a]$ be the equivalence class of $a \in S$. Show that $[a] \subseteq S$.

17.23   Below are some relations over the set $\{a, b, c, d, e\}$. For each of these, decide whether it is an equivalence relation, and in that case, find the equivalence class $[a]$ of a:

(a) $\{ \langle a, a \rangle, \langle b, b \rangle, \langle c, c \rangle, \langle d, d \rangle \}$
(b) $\{ \langle a, a \rangle, \langle b, b \rangle, \langle c, c \rangle, \langle d, d \rangle, \langle e, e \rangle, \langle a, b \rangle, \langle b, a \rangle \}$
(c) $\{ \langle a, a \rangle, \langle b, b \rangle, \langle c, c \rangle, \langle d, d \rangle, \langle e, e \rangle, \langle b, c \rangle, \langle b, d \rangle, \langle c, b \rangle, \langle d, b \rangle \}$
(d) $\{ \langle a, a \rangle, \langle b, b \rangle, \langle c, c \rangle, \langle d, d \rangle, \langle e, e \rangle, \langle a, c \rangle, \langle b, c \rangle \}$
(e) $\{ \langle a, a \rangle, \langle b, b \rangle, \langle c, c \rangle, \langle d, d \rangle, \langle e, e \rangle, \langle b, d \rangle, \langle d, b \rangle \}$
(f) $\{ \langle a, a \rangle, \langle b, b \rangle, \langle c, c \rangle, \langle d, d \rangle, \langle e, e \rangle, \langle b, d \rangle, \langle d, b \rangle, \langle a, c \rangle, \langle c, a \rangle \}$

17.24   Let $\sim$ be an equivalence relation on the natural numbers, and let E be $[0]$, that is, the equivalence class of the number 0.

(a) Prove that E does *not* equal the empty set.
(b) Prove that for all natural numbers x and y in E, we have that $x \sim y$.
(c) We have learned that if $x \sim y$, then $[x] = [y]$. Is the converse also the case, that is, that if $[x] = [y]$, then $x \sim y$? If yes, provide a proof; if no, provide a counterexample.

17.25   Suppose that $\sim$ is an equivalence relation on $\{A, B, C, D, E\}$ that gives rise to the equivalence classes $\{B, E\}$ and $\{A, C, D\}$. Write down $\sim$ as a set of ordered pairs.

17.26   We have defined the relation $\curvearrowright$ on bit strings by saying that $(s \curvearrowright t)$ holds exactly when t is obtained from s by moving the leftmost character such that it ends up being the rightmost character.

(a) Show that $\circlearrowleft$, which is the transitive closure of $\curvearrowright$, is an equivalence relation.
(b) Is it possible that $\curvearrowright$ can be an equivalence relation? If so, how can this be?
(c) What are the equivalence classes of 1234 and 1324? Are they identical?

Chapter 18

Combinatorics

■■■■■■■■■■■■■■■■■■■■■■■

In this chapter you will learn basic combinatorics. We will go through some basic counting principles, such as the inclusion–exclusion principle and the multiplication principle, and we will define permutations, ordered selections, combinations, and binomial coefficients.

The Art of Counting

Combinatorics is about counting and finding answers to questions of the form "in how many ways can we ...." Combinatorics is the study of enumerations, combinations, and permutations, and is important in, among other things, probability theory and complexity analysis of algorithms. In this chapter we are going to look at the basics of combinatorics, and first we will look at a few important principles for calculating the number of possibilities.

The Inclusion–Exclusion Principle

Definition 18.1. The inclusion–exclusion principle for two sets

When A and B are two finite sets, the **inclusion–exclusion principle** says that:

$$|A \cup B| = |A| + |B| - |A \cap B|$$

If we first count the elements of A and then the elements of B, we have counted the elements of $A \cap B$ twice. In order to get the number of elements of $A \cup B$, we must therefore subtract what we have overcounted, namely, the number of elements of $A \cap B$:

$$|A \cup B| \qquad |A| \qquad |B| \qquad |A \cap B|$$

© Springer Nature Switzerland AG 2021
R. Antonsen, *Logical Methods*, https://doi.org/10.1007/978-3-030-63777-4_19

Q **The inclusion–exclusion principle.** Let A be $\{a, b\}$ and let B be $\{b, c, d\}$. Then A $\cup$ B $= \{a, b, c, d\}$ and A $\cap$ B $= \{b\}$. We see that the principle is correct:

$$|A \cup B| = |A| + |B| - |A \cap B| = 2 + 3 - 1 = 4.$$

◆

Definition 18.2. The inclusion–exclusion principle for three sets

If A, B, and C are three finite sets, the **inclusion–exclusion principle** says that

$$|A \cup B \cup C| = |A| + |B| + |C| - (|A \cap B| + |A \cap C| + |B \cap C|) + |A \cap B \cap C|.$$

If we add together the numbers of elements of A, B, and C, we have counted the elements of A $\cap$ B and A $\cap$ C and B $\cap$ C twice, and we have counted the elements of A $\cap$ B $\cap$ C three times. Therefore we have to subtract $|A \cap B| + |A \cap C| + |B \cap C|$, but then we have subtracted the elements of A $\cap$ B $\cap$ C three times. By adding them back once, we get the number of elements of A $\cup$ B $\cup$ C. This can be illustrated in the following way:

$$|A \cup B \cup C| \qquad |A| \qquad |B| \qquad |C|$$

$$|B \cap C| \quad |A \cap C| \quad |A \cap B| \qquad |A \cap B \cap C|$$

The inclusion–exclusion principle may readily be generalized to more than three sets, and Venn diagrams are tools to keep track of which sets we have counted and the number of times for each.

The Multiplication Principle

The next principle for the calculation of the number of possibilities is the multiplication principle:

Definition 18.3. The multiplication principle

If we make a series of independent choices, the total number of possibilities is the product of the number of possibilities of each choice. This is called the **multiplication principle.**

This principle can be used to calculate the size of a Cartesian product:

$$|A_1 \times A_2 \times \cdots \times A_n| = |A_1| \cdot |A_2| \cdots |A_n|.$$

The number of elements of the Cartesian product equals the number of elements of each set multiplied together. For two sets A and B, we get $|A \times B| = |A| \cdot |B|$.

Q The multiplication principle, grade distribution. Suppose that two hundred students each get one out of six possible grades. The number of possible assignments of grades equals

$$\underbrace{6 \cdot 6 \cdot 6 \cdots 6}_{200 \text{ times}} = 6^{200}.$$

♦

Q The multiplication principle, on strings. *How many bit strings of length 5 are there?* In each position there can be either 0 or 1, and these are independent of each other. We can illustrate this by drawing a *tree* over all the possibilities. At the bottom of the tree, we choose the first character, then we choose the second character, and so on. In this way, each branch will correspond to exactly one string:

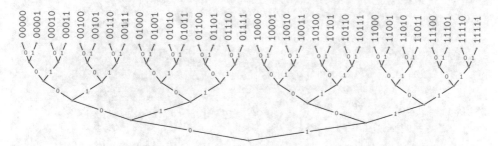

The multiplication principle gives that the total number of possibilities is the product of the number of possibilities for each choice. The total number of possibilities is

$$2 \cdot 2 \cdot 2 \cdot 2 \cdot 2 = 2^5 = 32.$$

How many strings of length 5 are there over the alphabet $\{0, 1, 2\}$? For each position there are three possibilities, and these are independent of each other. By drawing a tree similar to the one above, we get the following picture. Here there are *three* possibilities for each level:

The multiplication principle gives that the total number of possibilities is the product of the number of possibilities for each choice. The total number of possibilities is

$$3 \cdot 3 \cdot 3 \cdot 3 \cdot 3 = 3^5 = 243.$$

♦

Digression

How many relations are there on a set of three elements? A relation on a set A is a subset of the Cartesian product A × A. This set has nine elements, and each of these elements is either in the relation or not. This means there are $2^9 = 512$ possibilities in total. Here is an attempt to visualize all these relations in one and the same image. Although a set with only three elements is very simple, we get a large number of relations and a great deal of complexity.

- The 64 reflexive relations are marked with ⠒.
- The 64 symmetric relations are marked with ⠇.
- The 171 transitive relations are marked with □.
- The 27 functions are marked with ▮.
- The 27 relations whose *inverses* are functions, are marked with ⠆.

Notice that the top left relation is the *empty* relation and that bottom right relation is the *universal* relation. *Do you see the identity function? Do you see the six functions?*

Q   **The multiplication principle.** In the following diagram, one path from the left to the right side is marked. *How many such paths are there in total?*

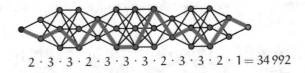

$$2 \cdot 3 \cdot 3 \cdot 2 \cdot 3 \cdot 3 \cdot 3 \cdot 2 \cdot 3 \cdot 3 \cdot 2 \cdot 1 = 34\,992$$

♦

If we imagine that we go from left to right and count the possibilities for each step, we get a series of choices that are *independent* of each other. That means that the total number of paths through the figure is the product of all the numbers, namely 34 992.

Permutations

A permutation is another word for a change of order, an ordering, or a shuffling. The word "permutation" comes from Latin and means a complete change (*permutatio*) or to change something completely (*permutare*). For example, when we shuffle a deck of cards, we are *permuting* the cards.

> **Definition 18.4. Permutation**
>
> A **permutation** of a set is an *ordering* of its elements. If we already have an ordering, a permutation is a change of the ordering.

Q **Permutations.** There are two ways to order the set $\{1, 2\}$: 12, and 21, and there are six ways to order the set $\{1, 2, 3\}$: 123, 132, 213, 231, 312, and 321.   ◆

We can define a permutation even more precisely by saying that a permutation of the set M is a *bijection* from M to M. There are, for example, six bijections from $\{1, 2, 3\}$ to $\{1, 2, 3\}$:

We see that these correspond to the six ways we ordered the set $\{1, 2, 3\}$ in the example. A good thing about this definition is that it also works for infinite sets. A *permutation of the integers* is a bijective function $f : \mathbb{Z} \to \mathbb{Z}$.

We have now reached the combinatorial question: *In how many different ways can we arrange n elements?* We have n possibilities for the first element, $(n - 1)$ possibilities for the second element, $(n - 2)$ possibilities for the third element, etc. This means that there are $n(n-1)$ ways to choose two elements in order, $n(n-1)(n-2)$ ways to choose three elements in order, and $n(n-1)(n-2) \cdots 1$ ways to choose n elements in order. To take the product of all the natural numbers from n to 1 is so common that it is abbreviated $n!$, which is defined in the following way:

> **Definition 18.5. The factorial function**
>
> If n is a natural number, then
>
> $$n! = n \cdot (n - 1) \cdot (n - 2) \cdots 1.$$
>
> We let $0! = 1$, and we read $n!$ as "n factorial."

In general, there are n! permutations of a set with n elements. Notice that 0! = 1. There are many good reasons for that. One good reason is that there is only one permutation of the empty set, corresponding to the empty function. Another good reason is how we defined the factorial function *recursively* in Chapter 10 (*page 114*).

Q **Permutations.** There are $4 \cdot 3 \cdot 2 \cdot 1 = 24$ permutations of the set $\{1, 2, 3, 4\}$. There are four possibilities for the first element, and for each of these possibilities, there are six permutations:

- 1 as the first element: 1234, 1243, 1324, 1342, 1423, 1432
- 2 as the first element: 2134, 2143, 2314, 2341, 2413, 2431
- 3 as the first element: 3124, 3142, 3214, 3241, 3412, 3421
- 4 as the first element: 4123, 4132, 4213, 4231, 4312, 4321

Each of these permutations corresponds to exactly one of the 24 bijections on $\{1, 2, 3, 4\}$:

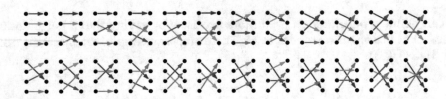

See whether you find any patterns in this figure. Do you see the identity function? Is this set closed under function composition?   ♦

We can see how the number of permutations grows by drawing trees that visualize all of the permutations:

n	n!	*visualization of all the permutations as a tree*
1	1	
2	2	
3	6	
4	24	
5	120	

Each tree in the table represents all the permutations on a set with n elements. Imagine starting at the bottom of each tree and going up toward the top. Each time you go one step up, you lose one option, and the branches become fewer. In the bottommost tree, for example, you have five branches to choose from when you are at the bottom; then the branches become fewer and fewer until you are at the top. Each *path through the tree* thus corresponds to one permutation.

Ordered Selection

Definition 18.6. Ordered selection, k-permutation

If a set of n elements is given, and we want to choose k of these in order, there are $n(n-1)(n-2)\cdots(n-(k-1))$ ways to do this. This is called an **ordered selection**, or a k-**permutation**, of n elements.

❷ In how many ways can we choose three elements in order from the set $\{1,2,3,4,5,6,7,8,9,10\}$? ◆

❶ The set has ten elements, and so there are $10\cdot 9\cdot 8 = 720$ ways to choose three elements in order. ◆

Notice that we can express $10\cdot 9\cdot 8$ using the factorial function in the following way:

$$10\cdot 9\cdot 8 = \frac{10\cdot 9\cdot 8\cdot 7\cdot 6\cdot 5\cdot 4\cdot 3\cdot 2\cdot 1}{7\cdot 6\cdot 5\cdot 4\cdot 3\cdot 2\cdot 1} = \frac{10!}{7!}.$$

Notation. A notation for this "falling factorial function" is

$$^{n}P_{k} = \frac{n!}{(n-k)!}$$

where "P" is for "permutation." Another notation for this number is $n^{\underline{k}}$. This is the product of k numbers, where the first number is n and where each number is one less than the previous one. The number $10\cdot 9\cdot 8$ can be written with this notation as $^{10}P_{3}$ or $10^{\underline{3}}$.

Q **Ordered selection.** There are $^{52}P_{2} = 52\cdot 51 = 2652$ ways to draw two cards in succession from a deck of 52 cards. ◆

The number $^{n}P_{k}$ tells us how many ways we can choose k elements in order from a set of n elements. If we look at the case $n = k$, we get the following, where we use that $0! = 1$.

$$^{n}P_{n} = \frac{n!}{(n-n)!} = \frac{n!}{0!} = \frac{n!}{1} = n!.$$

This is as expected, because there are n! permutations of a set of n elements.

Combinations

When we talk about permutations, we are discussing the number of ways in which we can choose elements *in order*, but we are often not interested in the order.

Definition 18.7. Combination

A **combination** is a selection of elements from a set where the order does not matter. A k-**combination** of a set A is a subset of A with k elements.

Q Combinations. Suppose we have five boxes, ☐☐☐☐☐, and that we are going to color exactly three of them. There are ten ways of doing this, and those are the following.

11100		$\{1,2,3\}$
11010		$\{1,2,4\}$
11001		$\{1,2,5\}$
10110		$\{1,3,4\}$
10101		$\{1,3,5\}$
10011		$\{1,4,5\}$

01110		$\{2,3,4\}$
01101		$\{2,3,5\}$
01011		$\{2,4,5\}$
00111		$\{3,4,5\}$

This is equivalent to looking at bit strings of length 5 with exactly three 1's, and each of these combinations corresponds to choosing a subset of $\{1,2,3,4,5\}$ with three elements. We have seen that the number of ways to choose three elements in *order* is $^5P_3 = 5 \cdot 4 \cdot 3 = 60$, but then each of the ten combinations shown above is counted $3! = 6$ times, which is the number of permutations of a set of three elements. If we are to take this into account, we must divide by 6 in order to get the number of combinations.   ♦

There are several notations for describing the number of k-combinations of a set of n elements. One of these is ^nC_k, where "C" comes from the word "combination," and another is $\binom{n}{k}$, which is read "n choose k." The number $\binom{n}{k}$ denotes the number of k-element subsets of a set of n elements.

We can generalize from the example above and describe the connection like this:

$$^nP_k = \binom{n}{k} k!$$

This holds because the number of ways to choose k elements in order is equal to the number of ways to choose k elements, regardless of order, multiplied by the number of permutations of these k elements. For example, we have that

$$^5P_3 = \binom{5}{3} 3! = 60.$$

If we divide by k! on both sides of the equation, we get:

$$\binom{n}{k} = \frac{^nP_k}{k!} = \frac{n!}{(n-k)!k!}.$$

We can check that it is correct for the numbers in the example:

$$\binom{5}{3} = \frac{5!}{(5-3)!3!} = \frac{5!}{2!3!} = \frac{5 \cdot 4 \cdot 3}{3 \cdot 2 \cdot 1} = 10.$$

This can be used to give a precise definition of the numbers $\binom{n}{k}$.

Definition 18.8. Binomial coefficient

If n and k are natural numbers such that $k \leqslant n$, we define $\binom{n}{k}$ by

$$\binom{n}{k} = \frac{n!}{(n-k)!k!}.$$

We read $\binom{n}{k}$ as "n choose k," and such a number is called a **binomial coefficient**.

Notice that

$$\binom{n}{k} = \binom{n}{n-k}$$

for all n and k. For example, $\binom{10}{7}$ equals $\binom{10}{3}$, because there are just as many subsets of size 7 as there are subsets of size 3. For each subset X with k elements, there is exactly one set with $n-k$ elements, namely $\overline{X}$. Said another way, there are just as many ways to choose k elements as there are ways to *choose away* k elements.

Repetitions and Overcounting

Sometimes we overcount, and then we must *compensate* for it in the right way. We have already seen this in the inclusion–exclusion principle, where we counted the elements of the *intersection* twice and had to subtract this once in order to get the correct number. But we can also overcount when we count permutations. Let us look at an example.

Q **Overcounting.** *How many different strings can we get by shuffling the characters of the string* koko? There are $4! = 24$ permutations of the string, because there are four characters in it, but many of the permutations give rise to identical strings. Let us pretend that the four characters are different – that it says koko with two different k's and two different o's – and look at all the permutations:

kkoo	k<u>k</u>oo	kk<u>oo</u>	k<u>k</u><u>oo</u>	<u>kk</u><u>oo</u>
koko	ko<u>k</u>o	k<u>o</u>ko	ko<u>k</u>o	k<u>o</u>ko
kook	ko<u>o</u>k	ko<u>o</u>k	ko<u>o</u>k	k<u>oo</u>k
okko	ok<u>k</u>o	ok<u>k</u>o	<u>o</u>kko	ok<u>k</u>o
okok	ok<u>ok</u>	okok	okok	okok
ookk	o<u>o</u>kk	o<u>o</u>kk	o<u>o</u>kk	<u>oo</u><u>kk</u>

Here we see to the left of the line that there are only six different strings that can occur, and to the right of the line we see how many times each of these occurs. Each string is counted *four* times. *Why is that so?* Look at the string kkoo. There are two ways to internally shuffle the k's, and there are two ways to internally shuffle the o's. Because these are independent of each other, the *multiplication principle* says that there are *four* ways to internally shuffle the k's and o's. In this case these are kkoo, kkoo, kkoo, and kkoo.

Let us try again. *How many different strings we can get by shuffling the characters of the string* pappa? Here there are *five* characters, two a's and three p's. There are $2! = 2$ ways to internally shuffle the a's and $3! = 6$ ways to internally shuffle the p's. That means that each *unique* word is counted $2! \cdot 3! = 12$ times:

$$\begin{array}{cccccc}
\overset{1\ 23}{\text{pappa}} & \overset{2\ 13}{\text{pappa}} & \overset{1\ 32}{\text{pappa}} & \overset{2\ 31}{\text{pappa}} & \overset{3\ 12}{\text{pappa}} & \overset{3\ 21}{\text{pappa}} \\
\overset{1\ 23}{\text{pappa}} & \overset{2\ 13}{\text{pappa}} & \overset{1\ 32}{\text{pappa}} & \overset{2\ 31}{\text{pappa}} & \overset{3\ 12}{\text{pappa}} & \overset{3\ 21}{\text{pappa}}
\end{array}$$

We therefore have to divide the number of permutations of five characters, which is 5!, by this number. We get $\frac{5!}{2! \cdot 3!} = 10$ possible strings. These ten are aappp, apapp, appap, apppa, paapp, papap, pappa, ppaap, ppapa, and pppaa. In both of these cases the answers are actually binomial coefficients, $\binom{4}{2}$ and $\binom{5}{3}$, respectively. We could just as well have formulated the questions this way: *Given four characters, in how may ways can we choose two of them?* and *Given five characters, in how may ways can we choose three of them?* ◆

In general, if we have n objects, with n_1 of one type, n_2 of another type, ..., and n_r of a last type – and we cannot distinguish between objects of the same type – there are

$$\frac{n!}{n_1! \cdot n_2! \cdots n_r!}$$

unique permutations of these n objects.

Exercises

18.1 (a) Suppose that we have four squares and that each square is colored either white or blue. Here are some of the possibilities: 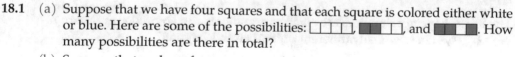. How many possibilities are there in total?

(b) Suppose that we have four squares and that each square is colored either white, blue, or green. Here are some of the possibilities: ▢▮▮▮, ▮▮▢▮, and ▮▮▮▮. How many possibilities are there in total?

(c) Suppose that we have eight squares, n colors, and that each square is colored differently. In how many ways can it be done?

(d) Suppose that we have eight squares and that we are coloring exactly two squares blue and exactly two squares green. Here are some ways to do it: ▢▮▮▢▢▢▮▮, ▮▮▮▮▢▢▢▢ and ▢▢▢▢▮▮▮▮. In how many ways can it be done?

(e) Suppose that we have eight squares and that we are coloring exactly two squares blue, *but they cannot be next to each other*. Here are some ways to do it: ▮▢▮▢▢▢▢▢, ▮▢▢▢▮▢▢▢, and ▢▢▮▢▢▢▮▢. In how many ways can it be done? Can you find a general formula for this, where there are n squares?

18.2 What is wrong with the following reasoning: "Suppose that two hundred students each receive a grade out of six possible grades. Then there are $200 \cdot 6 = 1200$ possible assignments of grades."

18.3 In a room with 23 programmers, there are 13 that know Java and 7 that know Python. If there are 5 who know both programming languages, how many are there who know neither?

18.4 (a) Find all the permutations of the set $\{1, 2\}$.

(b) How mange permutations are there of the set $\{a, b, c, d, e, f\}$?

18.5 Calculate the following expressions:

(a) $^{6}P_3$ (c) $^{6}P_1$ (e) $^{5}P_2$ (g) $\binom{3}{0}$ (i) $^{10}C_1$ (k) $\binom{12}{4}$

(b) $^{6}P_5$ (d) $^{9}P_1$ (f) $^{8}P_3$ (h) $\binom{4}{4}$ (j) $^{10}C_3$ (l) $\binom{15}{11}$

18.6 (a) How many groups of size four are there in a collection of seven people?

(b) Take a set of ten elements. How many subsets with *three elements* does it have?

18.7 Suppose we have a set with an even number of elements. In how many ways can this be divided up into two equally large and disjoint subsets? For example, the set $\{1, 2\}$ can only be divided up in one way: $\{\{1\}, \{2\}\}$. The set $\{1, 2, 3, 4\}$ can be divided up in three ways: $\{\{1, 2\}, \{3, 4\}\}$, $\{\{1, 3\}, \{2, 4\}\}$, and $\{\{1, 4\}, \{2, 3\}\}$. Suppose the set has $2n$ elements and find a general formula.

18.8 How many passwords can we create with the following restrictions? A password must have 8 characters, the first character must be a letter, and the rest must be letters *or* digits. We do not distinguish between uppercase and lowercase letters. A password must contain at least one digit.

18.9 The following eight numbers are given as arguments to a computer program: 3, 7, 4, 1, 2, 3, 4, 4. In how many different orders can this happen?

18.10 (a) How many different strings can we get by shuffling the characters in the string LOLOLOL?

(b) How many different strings can we create using one 0, one 1, two 2's, and three 3's? Briefly explain how you calculate the answer.

18.11 Let L be the language $\{a, b, c\}^*$, which consists of all strings over the alphabet $\{a, b, c\}$.

(a) How many strings in L have length 4?

(b) How many strings in L have length 5?

(c) How many strings in L of length 5 have at least two occurrences of a?

(d) How many strings in L of length 5 are *sorted*, meaning that all the a's are to the left of all the b's, and all the b's are to the left of all the c's?

18.12 (a) How many functions are there on a set of n elements?

(b) How many of these are bijective?

18.13 How many binary relations are there on a set with

(a) no elements?   (b) one element?   (c) two elements?

18.14 In each of the following diagrams, one path from left to right is highlighted. *How many such paths are there for each diagram?*

   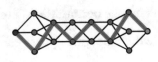

Chapter 19

A Little More Combinatorics

*In this chapter you will learn a little more about combinatorics. We will talk
more about the binomial coefficients, and go through an example in detail
that leads us to Pascal's triangle. Finally, we will look at how enumeration
problems can be systematized.*

Pólya's Example and Pascal's Triangle

The following figures show four different "paths" that lead from the top to the
bottom through a grid. Each path consists of eight steps: four left steps and four
right steps. *How many such paths are there?*

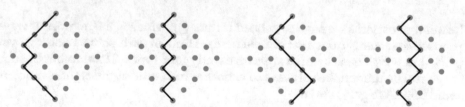

One way to approach this problem is by looking at variants of the problem that
are easier to handle. For example, we can count how many paths lead from the top
dot to other dots. For all the dots on the top outer edge, there is only one path. We
mark this with 1 in the left-hand figure below. There are exactly two paths that lead
from the top to the dot ● and there are exactly three paths that lead from the top to
the dot ●. If we continue like this, on both the left and the right, we can fill in the
numbers of paths that lead to each dot at the second level (in the lower right-hand
figure).

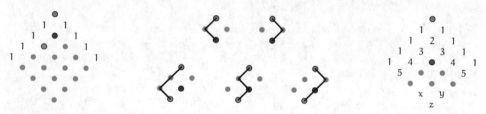

We eventually see that there is a pattern in the number of paths: *The number of paths,
z, to a dot equals the sum of the numbers of paths, x and y, to the dots right above*. In the

© Springer Nature Switzerland AG 2021
R. Antonsen, *Logical Methods*, https://doi.org/10.1007/978-3-030-63777-4_20

right-hand figure above, there are $3 + 3 = 6$ paths to the dot •. We double-check that this is correct by drawing all the paths:

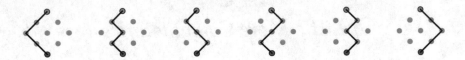

If we fill in the numbers, we see that there are 70 paths through the grid, each consisting of four left steps and four right steps. The numbers we have calculated are called binomial coefficients, and if we draw all the numbers that we can obtain in this way, we get **Pascal's triangle**, named after the French mathematician and philosopher *Blaise Pascal* (1623–1662):

$$
\begin{array}{ccccccccccccccccc}
&&&&&&&& 1 \\
&&&&&&& 1 && 1 \\
&&&&&& 1 && 2 && 1 \\
&&&&& 1 && 3 && 3 && 1 \\
&&&& 1 && 4 && 6 && 4 && 1 \\
&&& 1 && 5 && 10 && 10 && 5 && 1 \\
&& 1 && 6 && 15 && 20 && 15 && 6 && 1 \\
& 1 && 7 && 21 && 35 && 35 && 21 && 7 && 1 \\
1 && 8 && 28 && 56 && 70 && 56 && 28 && 8 && 1
\end{array}
$$

$$
\begin{array}{ccccccccccccccccc}
&&&&&&&& \binom{0}{0} \\
&&&&&&& \binom{1}{0} && \binom{1}{1} \\
&&&&&& \binom{2}{0} && \binom{2}{1} && \binom{2}{2} \\
&&&&& \binom{3}{0} && \binom{3}{1} && \binom{3}{2} && \binom{3}{3} \\
&&&& \binom{4}{0} && \binom{4}{1} && \binom{4}{2} && \binom{4}{3} && \binom{4}{4} \\
&&& \binom{5}{0} && \binom{5}{1} && \binom{5}{2} && \binom{5}{3} && \binom{5}{4} && \binom{5}{5} \\
&& \binom{6}{0} && \binom{6}{1} && \binom{6}{2} && \binom{6}{3} && \binom{6}{4} && \binom{6}{5} && \binom{6}{6} \\
& \binom{7}{0} && \binom{7}{1} && \binom{7}{2} && \binom{7}{3} && \binom{7}{4} && \binom{7}{5} && \binom{7}{6} && \binom{7}{7} \\
\binom{8}{0} && \binom{8}{1} && \binom{8}{2} && \binom{8}{3} && \binom{8}{4} && \binom{8}{5} && \binom{8}{6} && \binom{8}{7} && \binom{8}{8}
\end{array}
$$

The number of paths we have calculated is thus $\binom{8}{4}$, which is the number of ways to choose 4 elements from a set of 8 elements. Think of each path as a string over $\{\,\nearrow, \searrow\,\}$ in which each symbol appears exactly four times. The number of such strings equals the number of ways to choose which four of the eight characters should be $\nearrow$. We get $\binom{8}{4} = 70$.

> ### Digression
>
> The example with the grid was used in 1978 by *George Pólya* when at age 91 he taught combinatorics at Stanford University. Many famous number sequences are hiding in Pascal's triangle. See if you can find:
>
> - the triangular numbers $(1, 3, 6, 10, 15, 21, \ldots)$,
> - the powers of two $(1, 2, 4, 8, 16, 32, \ldots)$,
> - the square numbers $(1, 4, 9, 16, 25, \ldots)$,
> - the Fibonacci numbers $(1, 1, 2, 3, 5, 8, 13, 21, \ldots)$.
>
> Hint: Sometimes you have to add numbers. The American mathematician *David Singmaster* (1939–), who is known for his notation system for Rubik's cube (*page 228*), asked in 1971 whether there is any upper limit to how many times a number other than 1 occurs in Pascal's triangle. The number 2 occurs once, the numbers 3, 4, and 5 twice, and the number 6 three times. Singmaster proved in 1975 that infinitely many numbers occur at least six times, but no one knows whether there is a number that occurs exactly five or seven times. One number has been found that occurs eight times, and that is 3003.

The fact that each number in Pascal's triangle equals the sum of the two numbers above means that it is possible to define the binomial coefficients recursively, in the following way:

$$\binom{n+1}{k} = \binom{n}{k-1} + \binom{n}{k}.$$

To make this more concrete, let us go back to the example in which we were coloring exactly three out of five boxes. Then we get

$$\binom{5}{3} = \binom{4}{2} + \binom{4}{3}.$$

If the first box is colored, then two of the four remaining boxes must be colored. There are $\binom{4}{2} = 6$ ways to do this. If the first box is *not* colored, then three of the four remaining boxes must be colored. There are $\binom{4}{3} = 4$ ways to do this.

Binomial Coefficients

Why are the numbers $\binom{n}{k}$ called binomial coefficients? In order to answer that, let us expand expressions of the form $(x+y)^n$. The following are all such expressions up to $n = 5$:

$$(x+y)^0 = 1,$$
$$(x+y)^1 = 1x^1 + 1y^1,$$
$$(x+y)^2 = 1x^2 + 2x^1y^1 + 1y^2,$$
$$(x+y)^3 = 1x^3 + 3x^2y^1 + 3x^1y^2 + 1y^3,$$
$$(x+y)^4 = 1x^4 + 4x^3y^1 + 6x^2y^2 + 4x^1y^3 + 1y^4,$$
$$(x+y)^5 = 1x^5 + 5x^4y^1 + 10x^3y^2 + 10x^2y^3 + 5x^1y^4 + 1y^5.$$

We see that all the coefficients, that is, the numbers in front of the expressions of the form $x^{n-k}y^k$, are the numbers $\binom{n}{k}$ from Pascal's triangle. These numbers are called "binomial coefficients," because expressions of the form $x + y$ are called *binomials*, which are *polynomials* in two variables. In general, we have that $(x+y)^n$ equals:

$$\binom{n}{0}x^ny^0 + \binom{n}{1}x^{n-1}y^1 + \binom{n}{2}x^{n-2}y^2 + \cdots + \binom{n}{n-1}x^1y^{n-1} + \binom{n}{n}x^0y^n.$$

We can see this by analyzing the expression:

$$(x+y)^n = \underbrace{(x+y)(x+y)\cdots(x+y)}_{n \text{ occurrences}}.$$

If we multiply this out, we get 2^n terms, and each of these terms is of the form $f_1 f_2 \cdots f_n$, where each factor f_i is either x or y. The question is how many of these terms are identical. The answer is that there are $\binom{n}{k}$ terms in which there are k occurrences of the factor y (and therefore $n - k$ occurrences of x). For example, if we expand $(x + y)^4$, we get $\binom{4}{2} = 6$ terms with 2 occurrences of y:

$$xxyy, xyxy, xyyx, yxxy, yxyx, yyxx.$$

All these equal x^2y^2, and this gives us that the coefficient of x^2y^2 is 6.

Digression

What is the largest number you can write down on a sheet of paper in thirty seconds? In combinatorics we quickly get to very *large numbers*. Using only a few characters, like 10^{80}, we can describe the number of atoms in the universe, but this number is microscopic compared to other numbers. The American mathematician and computer scientist *Donald Knuth* (1938–) introduced a notation in 1976, which today is just called "Knuth's arrow notation," for describing very large numbers. The idea behind the notation is that multiplication can be seen as repeated addition, exponentiation as repeated multiplication, etc.

$$
\begin{aligned}
a \uparrow b &= aa \cdots a & &\text{where } a \text{ occurs } b \text{ times} \\
a \uparrow\uparrow b &= a \uparrow (a \uparrow (\cdots \uparrow a)) & &\text{where } a \text{ occurs } b \text{ times} \\
a \uparrow\uparrow\uparrow b &= a \uparrow\uparrow (a \uparrow\uparrow (\cdots \uparrow\uparrow a)) & &\text{where } a \text{ occurs } b \text{ times} \\
a \uparrow^n b &= a \uparrow^{n-1} (a \uparrow^{n-1} (\cdots \uparrow^{n-1} a)) & &\text{where } a \text{ occurs } b \text{ times}
\end{aligned}
$$

For example, $2 \uparrow\uparrow 5 = 2 \uparrow (2 \uparrow (2 \uparrow (2 \uparrow 2))) = 2^{2^{2^{2^2}}} = 2^{2^{2^4}} = 2^{2^{16}} = 2^{65536}$.

A well-known number, called **Graham's number**, named after the American mathematician and juggler *Ronald Graham* (1935–2020), can be described using this notation: if $g_1 = 3 \uparrow\uparrow\uparrow\uparrow 3$ and $g_n = 3 \uparrow^{g_{n-1}} 3$, then Graham's number is g_{64}. This is one of the largest numbers used in a mathematical proof. And if you think this number is too small, the British mathematician *John Horton Conway* (1937–2020) created a notation for describing even larger numbers. A number in this notation is a sequence of numbers with arrows between them. For example, $a \to b \to n$ is a way of expressing $a \uparrow^n b$, and $3 \to 3 \to 3 \to 3$ is a number far greater than Graham's number. It is worth noting that all of these definitions are examples of recursively defined functions.

Systematization of Counting Problems

It is possible to classify counting problems in several ways. Some words and distinctions that recur in such classifications are *ordered/unordered selection* and *with/without repetition*. When we use words like these, we often have a specific model in mind, for example a finite set from which we visualize choosing elements. If we make an *ordered selection* from a set, the order matters; we can imagine that we label the elements according to the order in which they are chosen. If we make an *unordered selection*, we do not care about the order. In choosing, we can either return each element to the set after it has been chosen, which is referred to as selection *with repetition* or *replacement*, or not, which is referred to as selection *without repetition* or *replacement.* If we have a set of five elements and are choosing three elements, it can be done in the following ways:

	ordered selection	unordered selection
with repetition	$5^3 = 5 \cdot 5 \cdot 5 = 125$	$\binom{7}{3} = \frac{7 \cdot 6 \cdot 5}{3 \cdot 2 \cdot 1} = 35$
without repetition	$^5P_3 = 5 \cdot 4 \cdot 3 = 60$	$\binom{5}{3} = \frac{5 \cdot 4 \cdot 3}{3 \cdot 2 \cdot 1} = 10$

If we have a set of five elements and are making an *unordered selection* of three elements *with repetition*, we are interested only in *how many times each element is chosen*. We can imagine that we have five cubicles and that we are placing each of three stars in one or another cubicle. This will be the same as if we had 7 characters of which 3 must be stars (and thus 4 must be walls). Then we get $\binom{7}{3} = \frac{7 \cdot 6 \cdot 5}{3 \cdot 2 \cdot 1} = 35$ possibilities:

```
★★★||||   ★|★★|||   ★||★|★|   |★★★|||   |★|★|★|   ||★★★||   ||★||★★
★★|★|||   ★|★|★||   ★||★||★   |★★★|||   |★|★||★   ||★★★|★   |||★★★|
★★||★||   ★|★||★|   ★|||★★|   |★★||★|   |★||★★|   ||★★||★   |||★★|★
★★|||★|   ★|★|||★   ★|||★|★   |★★|||★   |★||★|★   ||★|★★|   |||★|★★
★★||||★   ★||★★||   ★||||★★   |★|★★||   |★|||★★   ||★|★|★   ||||★★★
```

In general, there are $\binom{n+k-1}{k}$ ways of making an unordered selection with repetition of k elements from an n-element set. If we are choosing k elements from a set of n elements, it can be done in the following ways:

	ordered selection	unordered selection
with repetition	n^k	$\binom{n+k-1}{k}$
without repetition	^nP_k	$\binom{n}{k}$

Another perspective, and another way to describe the same thing, is by counting the number of functions $f : K \to N$, where K has k elements and N has n elements. In the same way as above, we get four combinations: we can choose whether we count all functions or only the injective ones, and whether or not we identify the elements of K with each other and consider only the number of times each element of N is "hit." In the four different cases, it is the number of functions we count. We get the following table. Notice the similarity between this table and the previous one:

	do not identify K	identify K
all functions	n^k	$\binom{n+k-1}{k}$
injective functions	^nP_k	$\binom{n}{k}$

We can imagine that we have k balls that we are putting in n boxes. One function
f : K → N will assign a box f(b) to each ball b. Requiring a function to be injective
means that each box can hold a maximum of one ball. Without this requirement,
the boxes may contain several balls. Identifying elements of K with each other
means intuitively that it is impossible to distinguish between the balls. This can be
done precisely by introducing an *equivalence relation* between the functions. We can
illustrate this in the following way; in the figure to the left we see the difference
between the elements of K, but in the figure to the right we do not:

Digression

The table below shows the number of partitions B_n of a set of size n:

n	0	1	2	3	4	5	6	7	8	9	10
B_n	1	1	2	5	15	52	203	877	4140	21147	115975

These numbers are called **Bell numbers**, named after the British-American mathematician and science-fiction author *Eric Temple Bell* (1883–1960), who studied them
in the 1930s. There are several formulas that express these numbers, but instead, we
shall look at a *recursive* way of calculating them:

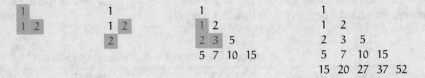

We create the table by beginning with two 1's one above the other. We take their
sum and write it to the right of the bottommost 1. Because there is no number *above*
this number, we copy it to *beginning* of the next row and continue: we always take
the sum of one number and the number just above it if there is a number there. The
sum of 1 and 2 is 3, and we write this to the right of 2. The sum of 2 and 3 is 5. But
now there is no number *above* 5, so we copy it to the beginning of the next row and
continue: the sum of 2 and 5 is 7. The sum of 3 and 7 is 10. We continue like this, and
the Bell numbers appear in the first and last positions of each row! The Bell numbers
are also related to the factorization of integers: B_n is the number of ways to factor a
number that is the product of n distinct primes. The number 30, for example, is the
product of three distinct primes, 2, 3, and 5, and can be factored in $B_3 = 5$ different
ways corresponding to the partitions of $\{2, 3, 5\}$, namely $2 \cdot 3 \cdot 5, 2 \cdot 15, 3 \cdot 10, 5 \cdot 6$, and
30. Some of the Bell numbers are prime. For example, 2, 5, and 877 are primes. These
are B_2, B_3, and B_7, and the next is $B_{13} = 27\,644\,437$. No one knows whether there are
infinitely many of them. Seven have been found, and the largest is B_{2841}, which is
approximately 10^{6539}.

Exercises

19.1 How many bit strings of length eight start with 1 or end with 00?

19.2 We say that a string is a **palindrome** if the result is the same whether it is read forward or backward. For example, abba and 1234321 are palindromes. How many palindromes of length m are there over an alphabet with n characters?

19.3 In how many ways can we define a set of two natural numbers both of which are less than 100?

19.4 Suppose that n is a positive integer.

  (a) How many functions are there from the set $\{1, 2, \ldots, n\}$ to $\{0, 1\}$?
  (b) How many functions are there from the set $\{1, 2, \ldots, n\}$ to $\{0, 1, 2\}$?
  (c) How many functions are there from the set $\{1, 2, \ldots, n\}$ to $\{0, 1, 2, 3\}$?
  (d) For each case above, how many of these are *injective*?

19.5 Let M be the set $\{1, 2, 3, 4, 5\}$.

  (a) How many subsets of M contain either 1, 2, or both?
  (b) How many functions from M to M are there, and how many of these are bijections?

19.6 Explain the following identity:

$$\binom{n}{k} + \binom{n}{k+1} + \binom{n}{k+2} + \binom{n}{k+3} = \binom{n+1}{k+1} + \binom{n+1}{k+3}.$$

19.7 In how many different ways can we go from A to B in this $2 \times 2 \times 2$ cube? We have to go one step at a time and only upward, to the right, or inward.

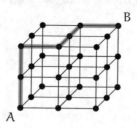

19.8 In how many different ways can six grades be assigned to ten students? We are not interested in what grade a particular student gets, but only the number of each grade in the distribution.

19.9 Suppose that the alphabet $\{0, 1, 2\}$ is given, and let us say that a string is **sorted** if the numbers are not decreasing in value as read from left to right. For example, 001122 and 002 are sorted, but 221100 and 201 are not.

(a) Find all sorted strings of length one, two, and three.

(b) How many sorted strings of length four are there? Of length five?

(c) How many sorted strings of length m are there?

(d) What is the answer to the previous questions if there are n characters in the alphabet?

19.10 Here we shall see that there is a connection between Morse code and Fibonacci numbers, as defined in Chapter 10 (*page 114*). Let a **Morse code** be a string over the alphabet { $\blacksquare$, $\rule[0.3em]{1em}{0.15em}$ }. Define the length $L(x)$ of a Morse code x recursively in the following way:

$$L(\Lambda) = 0, \qquad\qquad L(x\,\blacksquare) = L(x) + 1, \qquad\qquad L(x\,\rule[0.25em]{1em}{0.12em}) = L(x) + 2.$$

For example, we get the following lengths for some simple Morse codes:

$$L(\,\blacksquare\,\rule[0.25em]{1em}{0.12em}\,\blacksquare\,\blacksquare\,) = 5, \qquad L(\,\rule[0.25em]{1em}{0.12em}\,\rule[0.25em]{1em}{0.12em}\,\blacksquare\,) = 5, \qquad L(\,\blacksquare\,\blacksquare\,\blacksquare\,\blacksquare\,) = 3,$$
$$L(\,\rule[0.25em]{1em}{0.12em}\,\rule[0.25em]{1em}{0.12em}\,\rule[0.25em]{1em}{0.12em}\,) = 6, \qquad L(\,\blacksquare\,\blacksquare\,) = 2, \qquad L(\,\rule[0.25em]{1em}{0.12em}\,\blacksquare\,\rule[0.25em]{1em}{0.12em}\,) = 5.$$

Here are all Morse codes with lengths up to seven:

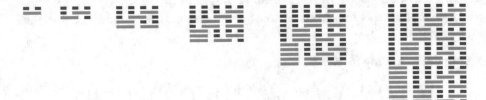

(a) Check that the numbers of Morse codes of length 2, 3, 4, 5, 6, and 7 are all Fibonacci numbers.

(b) Prove that the number of Morse codes of length n is a Fibonacci number.

Chapter 20

A Bit of Abstract Algebra

■·■■·■·■■■■■■■■■■■■■■·■·■■■

In this chapter you will learn some concepts from abstract algebra, in par-
ticular a bit more about relations and functions. We will talk about inverse
relations and functions, a few properties of operations, such as commutativ-
ity, associativity, and idempotency, and a few special properties of elements,
such as being an identity or inverse. Finally, we will look at what a group is.

Abstract Algebra

Algebra is a central part of mathematics, and *abstract algebra* is considered the part
of algebra that deals with so-called algebraic structures. An **algebraic structure**
is simply a set together with one or more operations on this set. We have already
seen several examples of operations, and this is just a new word for something we
already know. We will now look at some useful concepts from abstract algebra.

Inverse Relations and Functions

We studied relations and functions in Chapters 6 and 7, and we learned that a
relation is a set of tuples. We can always find the *opposite* relation by replacing each
tuple $\langle x, y \rangle$ with $\langle y, x \rangle$, or just *turn the arrows around* in a figure, and this is called
the *inverse* relation.

Definition 20.1. Inverse relation

If R is a relation from A to B, the **inverse relation** of R is the relation $\{ \langle y, x \rangle \mid \langle x, y \rangle \in R \}$ from B to A. We write R^{-1} for the inverse of R.

Q **Inverse relation.** The inverse relation of $\{ \langle a, 1 \rangle, \langle b, 2 \rangle \}$ is $\{ \langle 1, a \rangle, \langle 2, b \rangle \}$. If R is the
less-than relation on natural numbers, R^{-1} is the greater-than relation on natural
numbers. ◆

Each relation has an inverse, but it is not the case that we always get a function if we
take the inverse relation of a *function*. For example, the relation $\{ \langle a, 1 \rangle, \langle b, 2 \rangle, \langle c, 2 \rangle \}$

© Springer Nature Switzerland AG 2021
R. Antonsen, *Logical Methods*, https://doi.org/10.1007/978-3-030-63777-4_21

is a function from $\{a, b, c\}$ to $\{1, 2\}$, but the inverse relation, $\{\langle 1, a\rangle, \langle 2, b\rangle, \langle 2, c\rangle\}$, is not a function. *When does a function have an inverse?*

Definition 20.2. Inverse function

Let f be a one-to-one correspondence from the set A to the set B. The **inverse function** of f is the function from B to A such that $f^{-1}(b) = a$ if $f(a) = b$. We write f^{-1} for the inverse function of f.

It turns out that only one-to-one correspondences, or bijections, have inverse functions.

Q   Inverse function. Let f be a function from $\{1, 2, 3\}$ to $\{a, b, c\}$ such that $f(1) = a$, $f(2) = b$, and $f(3) = c$. *Does this function have an inverse?* Yes, it is a bijection. The inverse function f^{-1} is such that $f^{-1}(a) = 1$, $f^{-1}(b) = 2$, and $f^{-1}(c) = 3$.   ◆

Q   Inverse function. Let f be the function from integers to integers such that $f(x) = x+1$. *Does this function have an inverse?* Yes, because it is a one-to-one correspondence. The inverse function is in this case f^{-1} such that $f^{-1}(y) = y - 1$. For example, we have that $f(3) = 4$ and $f^{-1}(4) = 3$.   ◆

Some Properties of Operations

In the same way that we can study properties of sets, formulas, relations, and functions, the *operations* have some properties that are important to recognize. Here is a small selection of such properties.

Definition 20.3. Commutative

A binary operation $*$ on a set S is **commutative** if for all $x, y \in S$ it is the case that $x * y = y * x$.

The fact that a binary operation is commutative intuitively means that the order of the arguments does not matter. If you are putting coins into a machine, the order is usually unimportant.

Q   Commutativity. We already know several commutative operations:

- The operations $+$ (addition) and $\cdot$ (multiplication) are usually commutative, but the operations $-$ (subtraction) and $/$ (division on rational or real numbers without zero) are usually *not* commutative. A counterexample is, for example, the numbers 2 and 3, because $2 - 3 \neq 3 - 2$ and $2/3 \neq 3/2$.
- The operations $\cap$ (intersection) and $\cup$ (union) on sets are commutative, but the operation $\setminus$ (set difference) is *not* commutative.   ◆

Definition 20.4. Associative

A binary operation $*$ on a set S is **associative** if for all $x, y, z \in S$ it is the case that $x * (y * z) = (x * y) * z$.

That an operation is associative intuitively means that the placement of the parentheses does not matter.

Q Associativity. We already know several associative operations:

-   The operations $+$ (addition) and $\cdot$ (multiplication) are associative.
-   The operations $-$ (subtraction) and $/$ (division on rational or real numbers without zero) are *not* associative. A counterexample is, for example, the numbers $1, 2,$ and $3,$ because $2 = 1 - (2 - 3) \neq (1 - 2) - 3 = -4$ and $3/2 = 1/(2/3) \neq (1/2)/3 = 1/6$.
-   The operations $\cap$ (intersection) and $\cup$ (union) on sets are associative.
-   The operation $\setminus$ (set difference) is *not* associative.   ◆

The last the property we are going to look at is whether it has any effect to apply an operation more than once.

Definition 20.5. Idempotent

A unary operation f on a set S is **idempotent** if for all $x \in S$ it is the case that $f(f(x)) = f(x)$. A binary operation $*$ on a set S is **idempotent** if for all $x \in S$ it is the case that $x * x = x$.

The concept of *idempotency* is both natural and useful. The fact that something is idempotent means that it does not have any effect to do something more than once. Think of an activation button, for example at a crosswalk, where it does not help to press it more than once.

Q Idempotent. The absolute value function on real numbers is idempotent:

$$|x| = \begin{cases} x & \text{if } x \geqslant 0, \\ -x & \text{if } x < 0. \end{cases}$$

No matter what x is, we have $\|x\| = |x|$.   ◆

Q Idempotent. We have already seen several idempotent operations:

-   The operations $\cap$ (intersection) and $\cup$ (union) on sets are idempotent, because $X \cup X$ and $X \cap X$ always equal X, regardless of what X is.

– The operations + (addition) and · (multiplication) are *not* idempotent. A counterexample is the number 2, because $2 + 2 \neq 2$ and $2 \cdot 2 \neq 2$.   ◆

❷ Is / (division) idempotent?   ◆

❶ This is a trick exercise, because / is not an operation, and only operations can have the *idempotent* property. If we had turned it into an operation, for example by removing 0, it would still not have been idempotent, because $2/2 = 1$ is a counterexample.   ◆

Some Elements with Special Properties

It is also the case that when an operation on a set is given, there are some elements that behave in a special way. Some elements are, for example, *neutral* in the sense that they do not have an effect on the other elements.

Definition 20.6. Identity element

Let a binary operation ∗ on a set S be given. If $x * e = e * x = x$ for all $x \in S$, we say that e is an **identity element** or a **neutral element** for the operation ∗.

Q **Identity element.** For most sets of numbers, 0 is an identity element for addition, and 1 an identity element for multiplication. In set theory, $\varnothing$ is an identity element for union, and the universal set is an identity element for intersection. In formal languages, $\wedge$ is an identity element for concatenation. If we look at function composition, the identity function is an identity element.   ◆

Whether an element is an identity element thus depends on what happens in the interplay between this element and the other elements. It can also happen that some elements of a set neutralize each other, and we say that such elements have inverses.

Definition 20.7. Inverse elements

Let a binary operation ∗ on a set S be given, and suppose that e is an identity element for ∗. If there are elements $a, b \in S$ such that $a * b = b * a = e$, we say that a and b are **inverse** elements. We may also say that the **inverse** of a is b and write $a^{-1} = b$.

Notice that whether elements are inverses depends on the set, the operation, and the identity element e.

Q **Inverse elements.** It is not always the case that elements have inverses. If we look at the natural numbers and addition, only 0 has an inverse, and that is 0 itself: $0+0 = 0$. If we look at the natural numbers and multiplication, only 1 has an inverse, and that is 1 itself: $1 \cdot 1 = 1$. If we look at the integers and addition, all elements have inverses: the inverse of x is $-x$, because $x + (-x) = 0$. If we look at the rational numbers and multiplication, all elements except 0 have inverses: the inverse of x is $1/x$, because $x \cdot (1/x) = 1$.   ♦

Groups

There is a very special class of algebraic structures called *groups*. The following definition is an abstract way of characterizing all these structures.

Definition 20.8. Group

Let a binary operation • on a set G be given. Then $\langle G, \bullet \rangle$ is a **group** if the following conditions, called the **group axioms**, are satisfied.

- The operation • is associative.
- There is an identity element for •.
- Every element has an inverse.

If we additionally have that • is commutative, the group is called **abelian**.

Q **A group and a structure that is not a group.**

- $\langle \mathbb{Z}, + \rangle$, the integers under addition, is a group, because + is associative, 0 is an identity element, and each element x has an inverse $-x$. It is an *abelian* group because + is commutative.
- $\langle \mathbb{Z}, \cdot \rangle$, the integers under multiplication, is not a group, because not all elements have an inverse.   ♦

> **Digression**
>
> The concept of an *abelian group* is named after the Norwegian mathematician *Niels Henrik Abel* (1802–1829), who is considered one of the greatest mathematical talents in all of history, despite the fact that he died very young. He is famous, among other things, for having proved in 1823 that there is no general solution for fifth-degree equations using only so-called radical expressions, as we have for equations of degree less than five, like the quadratic formula. Since 2003, The Norwegian Academy of Science and Letters has annually awarded the **Abel Prize**, a math award of around 7.5 million Norwegian Kroner, to internationally renowned mathematicians.

Groups are used extensively in both computer science and physics as tools to represent and analyze different forms of *symmetries*. If we add more operations and axioms, we also get other interesting algebraic structures, like *rings*, *fields*, and *vector spaces*. In *group theory* we study the properties of groups and related structures.

Digression

Rubik's Cube was invented in 1974 by the Hungarian architect *Ernő Rubik* and is an algebraic structure in disguise. The cube consists of six faces, and on each face there are nine stickers, each attached to its own piece. One move consists in rotating one of the faces a quarter turn clockwise. It is common to name each move and face in this way:

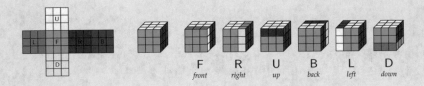

| F | R | U | B | L | D |
| front | right | up | back | left | down |

The cube can be analyzed and understood via group theory. We say that the surface of the cube consists of six center pieces, eight corner pieces, and twelve edge pieces. A sequence of moves causes the corner pieces and edge pieces to be repositioned, and this repositioning is called a *permutation* of the cube. If X and Y are two permutations, we can compose them to get a permutation XY, which consists in doing X followed by Y. The set of permutations together with this operation is a *group*, because (1) the operation is associative, (2) there is an identity element, and (3) all the permutations have inverses. The identity element is the permutation that does not move any of the pieces, and the inverse of a permutation X is the permutation X′ that corresponds to making the moves "backward," that is, in the opposite order and the opposite direction. This group is called the **cube group**, and many questions about the cube can be answered by studying it. For example, the cube group is not commutative, because the sequence of moves RU gives something different from the moves UR. If we perform a permutation X repeatedly, we will eventually return to the starting point. The set of permutations on this path back to the origin constitutes a so-called *cyclic subgroup* of the cube group. It is also possible to use combinatorics and group theory to calculate the number of permutations of the cube group:

$$\frac{8! \cdot 12! \cdot 3^8 \cdot 2^{12}}{2 \cdot 2 \cdot 3} = 43\,252\,003\,274\,489\,856\,000\,.$$

This means that the cube can be in over 43 billion billion different states. *What is the least number of moves needed to solve any cube?* This question has an interesting history, and in 2010 the answer was found: twenty moves are sufficient for solving any cube, and for some cubes twenty moves are necessary.

Exercises

20.1   For each of the following relations, find the inverse relation:

(a) $\{\langle 1,2 \rangle, \langle 1,3 \rangle, \langle 1,4 \rangle \}$
(b) $\{\langle 1,2 \rangle, \langle 2,1 \rangle, \langle 1,3 \rangle \}$
(c) The "is the daughter of" relation on the set of all humans.
(d) The "is a multiple of" relation on the set of positive integers.

20.2   Let $R = \{\langle 3,2 \rangle, \langle 2,3 \rangle, \langle 1, a \rangle \}$ and $f = \{\langle 1,2 \rangle, \langle 2,3 \rangle, \langle 3, a \rangle, \langle a, b \rangle, \langle b, 2 \rangle \}$.

(a) What is the inverse relation of R?
(b) Explain briefly why the function f does not have an inverse.

20.3   Let R be the relation $\{\langle a, b \rangle \mid a$ is strictly greater than $b\}$ on the set of integers.

(a) What is R^{-1}? (b) What is $\overline{R}$?

20.4   Let R be the relation $\{\langle a, b \rangle \mid b$ is divisible by $a\}$ on the set of integers.

(a) What is R^{-1}? (b) What is $\overline{R}$?

20.5   For each of the following functions, determine whether the function is a bijection, and if it is, find the inverse function:

(a) $\{\langle 0,0 \rangle, \langle 1,10 \rangle, \langle 2,20 \rangle, \langle 3,30 \rangle, \dots \}$ on the set of natural numbers
(b) $f(x) = 2x$ on the set of natural numbers
(c) $f(x) = 2x$ on the set of integers.
(d) $f(x) = 2x + 1$ on the set of real numbers
(e) $f(x) = x^2 + 1$ on the set of real numbers
(f) $f(x) = x^3$ on the set of real numbers

20.6   Let $f : \mathbb{R} \to \mathbb{R}$ be a function defined on the real numbers. We say that f is an *even function* if $f(-x) = f(x)$, and an *odd function* if $f(-x) = -f(x)$, for all $x \in \mathbb{R}$.

(a) Show that the function $f(x) = x^2$ is an even function.
(b) Show that the function $g(y) = y$ is an odd function.
(c) Suppose an odd function $f : \mathbb{R} \to \mathbb{R}$ is a bijection. Show that the inverse function f^{-1} is also an odd function.
(d) Why does an even function not have an inverse?

20.7   Find first-order formulas representing the following propositions:

(a) *The relation R is the inverse relation of S.*
(b) *The function g is the inverse function of f.*
(c) *The operation $\star$ is commutative.*
(d) *The operation $\star$ is associative.*

(e) *The unary operation f is idempotent.*

(f) *The binary operation ⋆ is idempotent.*

(g) *There is an identity element for ⋆.*

(h) *All elements have an inverse.*

20.8 (a) Prove by structural induction that the recursively defined function + on lists, as defined in Chapter 10 (*page 117*), is associative.

(b) Is it commutative?

(c) Is it idempotent?

20.9 What is wrong with the following statement? "We have that $f(2) = 2$. Therefore f is idempotent."

20.10 For each of the following choices for G and •, decide whether $\langle G, \bullet \rangle$ becomes a group.

(a) Let G be $\mathbb{N}$, and let $a \bullet b = a + b$

(b) Let G be $\mathbb{Z}$, and let $a \bullet b = a + b$

(c) Let G be $\mathbb{Z}$, and let $a \bullet b = a \cdot b$

(d) Let G be $\mathbb{R}$, and let $a \bullet b = a + b$

(e) Let G be $\mathbb{R} \setminus \{0\}$, and let $a \bullet b = a \cdot b$

(f) Let G be $\mathbb{R} \setminus \{0\}$, and let $a \bullet b = \frac{a}{b}$

20.11 Let M be a finite set, and let G be the set of all bijections on M. In Chapter 7 (*page 84*) we defined $g \circ f$ as the *composition* of the functions f and g.

(a) Show that $\circ$ is associative.

(b) Show that there is an identity element for $\circ$.

(c) Show that all elements in G have an inverse.

(d) Can we conclude that G is a group?

20.12 Create a group with four elements.

Chapter 21

Graph Theory

■■■■■■■■■■■■■■■■■■■■■■■■■■■■■■■

In this chapter you will learn basic graph theory. We will define graphs,
edges, and vertices, and we will introduce terminology to talk about and
categorize graphs in a precise way. You will learn a bit about directed graphs,
complete graphs, and the complement of graphs. You will also learn two
graph-theoretic results that apply to the vertices of a graph. Finally, we will
look at what it means for graphs to be isomorphic.

Graphs Are Everywhere

Graphs are everywhere around us, and graph theory is an important part of both
applied and theoretical mathematics. Graph theory can be used to model problems
in computer science, biology, social science, linguistics, physics, and many other
disciplines. The reason that graph theory is so useful is that many problems can
be represented by graphs. We have encountered the idea of representation several
times already: we can represent information using bit strings, and we can represent
a mathematical problem as another problem that is easier to solve. A good representation makes it possible to disregard what is irrelevant; we capture the essence.
That is exactly what happens in graph theory.

What Is a Graph?

A *graph* consists of *vertices* (•) and *edges* (——). Here are some graphs.

  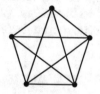 

Can you draw one or more of these graphs on a sheet without lifting the pencil and without
traversing any part of an edge more than once? With some graph theory, it is easy to
answer this question immediately. We shall see that this exercise is equivalent to
finding a so-called *Eulerian trail*.

© Springer Nature Switzerland AG 2021
R. Antonsen, *Logical Methods*, https://doi.org/10.1007/978-3-030-63777-4_22

Graphs as Representations

Graphs can represent many different things. The vertices can represent the students at a university, and the edges can represent acquaintanceship between students. The vertices can be the possible states of a computer program, and the edges can represent transitions between the states. A graph can represent the link structure of a website, where the vertices are web pages and the edges are links. The list goes on: electronic circuits, molecules in chemistry, computer networks, analysis of network traffic, and so on. Some graphs have special properties that are well suited for certain representations, for example *trees*. Let us look at some more examples of graphs before we start with the theory.

Q **Maps as graphs.** A map can provide a starting point for several different graphs, depending on what the vertices and the edges represent. One possibility is that the vertices represent *cities* and the edges represent *roads*. Another possibility is that the vertices represent *regions*, for example counties, and the edges represent *borders*. When we have the representation, we can disregard the original map:

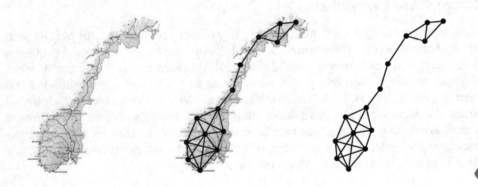

♦

Q **Road networks as graphs.** A road network can be represented as a graph. We can let each *intersection* correspond to a vertex, and we can let the *roads* connecting the intersections correspond to the edges. When we have drawn the graph, we can reason about it instead of the road network itself:

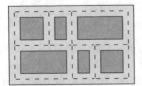 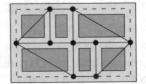 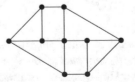

♦

Definitions and Concepts About Graphs

Definition 21.1. Graph

A **graph** G consists of a finite, nonempty set V of **vertices**, also called **nodes**, and a set E of **edges**. Each edge in E is a set $\{u, v\}$, where u and v are two distinct vertices. We say that $\{u, v\}$ connects u and v and that $\{u, v\}$ is **incident** to u and v. Two vertices are called **adjacent** if they are connected by an edge.

We normally draw edges as lines and vertices as dots, circles, or other symbols. But it is not important exactly how we draw the graph; it is the structure of a graph that is important, that is, which vertices are connected to which via an edge. For example, we do not care how long the edges are, what color they have, or whether they are bent. Here is the same graph drawn in four different ways. Here, A, B, C, and D are vertices, and e, f, g, and h are edges. The edge e is incident to the vertices A and B, and the vertices B and C are adjacent, because they are connected by the edge f.

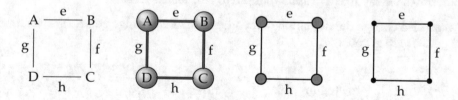

The fact that there is no unique way to draw a graph is not limited to how we draw the vertices and the edges. The following pairs of graphs have the same underlying structure, but are drawn in different ways. We can imagine that the edges are elastic and we can slide the vertices around. We will gradually clarify this through the concept of an *isomorphism*. Make sure you understand why the following graphs have the same structure:

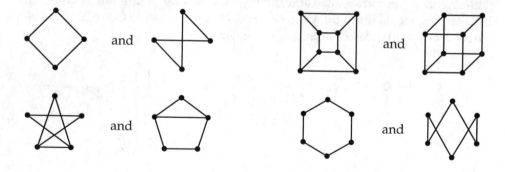

There are many other ways to define graphs. Therefore, when we read about graph theory, it is important to read the definitions carefully. By changing the definition of an edge, we get other concepts, and we will look at two of these. If an edge is defined such that multiple edges can be incident to the same vertices, or such that an edge can connect a vertex with itself, we get what are called **multigraphs** and **pseudographs**, respectively.

Definition 21.2. Loops and parallel edges

An edge from a vertex to itself in a pseudograph is called a **loop**. Two or more edges that connect the same vertices in a multigraph are called **parallel edges**. A multigraph or pseudograph is called **simple** if it has neither loops nor parallel edges.

It is common to define graphs such that neither loops nor parallel edges occur. Sometimes we say that a graph is simple to emphasize this.

We can also change the definition of an edge so that each edge gets a *direction*; then we get *directed* graphs.

Definition 21.3. Directed graph

A **directed graph** is defined as a graph with the additional condition that the edges are ordered pairs $\langle u, v \rangle$ instead of sets $\{u, v\}$.

The set of edges in a directed graph is exactly the same as a *binary relation*, which we encountered in Chapter 6. In the same way, we can say that the set of edges in a regular, *undirected*, graph is the same as a symmetric relation. In the same way as with relations, there are many equivalent ways to describe a directed graph: to the left as a diagram, in the middle as a set of ordered pairs, and to the right as a table, where each occurrence of 1 represents an edge:

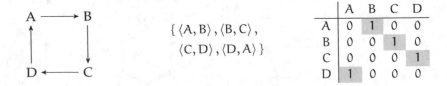

Q **Flowcharts as graphs.** We can think of a *flowchart* as a directed graph. In the following illustration, the vertices represent instructions in a computer program. *Can you see what the program does?*

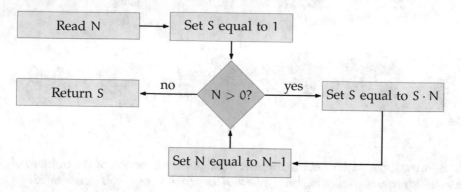

Properties of Graphs

We will first look at the smallest and the largest graphs.

Definition 21.4. Empty graph

A graph without edges is called an **empty graph** or a **null graph**.

The definition of a graph says that the set of vertices must be nonempty, but the set of edges can be empty.

Empty graphs.

Definition 21.5. Complete graph

A simple graph is **complete** if each vertex is adjacent to every other vertex. The complete graph with n vertices is denoted by K_n.

Q **Complete graphs.** Here are all the complete graphs from K_3 to K_{14}, and below each graph is the number of edges in the graph:

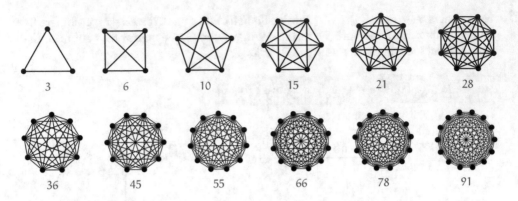

Try to figure out the pattern in the number of edges as a function of the number of vertices. *How many edges does* K_n *have? Which of the graphs have a "hole" in the middle?* ♦

Definition 21.6. Complement

If G is a graph, the **complement** of G is the graph that has the same vertices as G but in which two vertices are adjacent if and only if the vertices are not adjacent in G. We write $\overline{G}$ for the complement of G.

Here are some graphs and their complements:

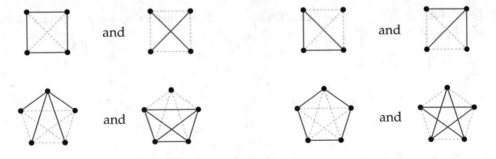

In the last the graph, at the bottom right, we do not get a new graph when we take the complement. The graph and its complement have the exact same underlying structure. Such graphs are called **self-complementary**.

Two Graph-Theoretic Results

Now we are going to look at two well-known results about graphs. Both have to do with the number of edges incident to a vertex.

Definition 21.7. The degree of a vertex

The **degree** of a vertex v is the number of edges that are incident to v. A loop counts as two edges. We denote the degree of v by $\deg(v)$. A vertex with degree 0 is called **isolated**.

Q   **The degree of a vertex.** A simple way to find the degree of a vertex is by drawing a circle around the vertex and counting how many edges intersect the circle. This method also works with loops and parallel edges, as in the following multigraph:

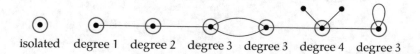

isolated   degree 1   degree 2   degree 3   degree 3   degree 4   degree 3

The first result is that the sum of the degrees of the vertices in a graph always equals twice the number of edges. If V is the set of vertices and E is the set of edges, this can be expressed in the following way:

$$\sum_{v \in V} \deg(v) = 2|E|.$$

We read this as "the sum of $\deg(v)$ over all v in V equals 2 times the number of elements of E." We can imagine that we start with just the vertices and then add the edges, one by one. For each edge we add, the sum of the degrees increases by two. If we add the degrees of all the vertices, each edge counts as two, because each edge is incident to two vertices.

Q   **Sum of the degrees.**

Each time we add an edge, the sum of the degrees increases by two:

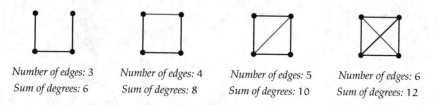

| Number of edges: 3 | Number of edges: 4 | Number of edges: 5 | Number of edges: 6 |
| Sum of degrees: 6 | Sum of degrees: 8 | Sum of degrees: 10 | Sum of degrees: 12 |

The second result is that every graph has an even number of vertices of odd degree. In Exercise 21.9 (*page 241*) you will be asked to prove this yourself. If we imagine that some people are gathered in a room and shaking hands, the number of people who shake hands with an odd number of people must be an even number. We can imagine that the vertices in a graph represent the people, and the edges represent those that are shaking hands. Because of this metaphor, this is often called the **handshaking lemma**. It is called a *lemma* because it mainly is a useful intermediate result for proving other lemmas and theorems.

Q **Number of vertices of odd degree.** Here are three graphs. The vertices with odd degree are marked with blue circles. Notice that there is always an even number of these:

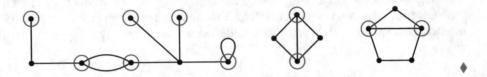

♦

Isomorphisms

The concept of an *isomorphism* is one of the most important in mathematics. The word comes from Greek and means *has the same shape*; "iso" means *equal* and "morph" means *shape*. Isomorphic objects have the same shape, but they can have different content. We say that two mathematical objects are isomorphic if they are "structurally identical" by certain criteria. If we have drawn two graphs and we can get from one to the other by sliding some vertices around, thereby changing the lengths and shapes of some edges, the graphs are isomorphic. The following graphs are all isomorphic:

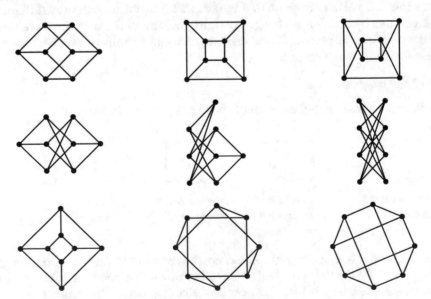

Imagine that the edges are rubber bands and that we can move the vertices around, but we are not allowed to add or remove vertices or edges. An isomorphism is a one-to-one correspondence that preserves certain properties. We shall now make this precise.

Definition 21.8. Isomorphism

Let G and H be two graphs. An **isomorphism** from G to H is a bijective function f from the vertices of G to the vertices of H such that the vertices u and v are adjacent in G if and only if the vertices f(u) and f(v) are adjacent in H.

Notation. It is common to write V(G) for the set of vertices in a graph G. When a graph is simple, it is common to write ab for an edge incident to the two vertices a and b. This does not work if the graph has parallel edges or loops.

Q **Isomorphism.** Let the graph G consist of the vertices {a, b, c, d} and the edges {ab, bc, cd, da}. Let the graph H consist of the vertices {1, 2, 3, 4} and the edges {12, 23, 34, 41}. The function f such that f(a) = 1, f(b) = 2, f(c) = 3, and f(d) = 4 is an isomorphism. For example, we see that a and b are adjacent in G, because ab is an edge. Then, f(a) and f(b), which are 1 and 2, must be adjacent in H. This is correct, because 12 is an edge in H:

Notice that the definition of isomorphism says "if and only if."

- The "if" part is that u and v are adjacent if the vertices f(u) and f(v) are. This means that if the vertices u and v are not adjacent, then neither are the vertices f(u) and f(v). In other words, adjacent in H implies adjacent in G.
- The "only if" part is that u and v are adjacent only if f(u) and f(v) are. This means that if the vertices u and v are adjacent, then the vertices f(u) and f(v) are also adjacent. In other words, adjacent in G implies adjacent in H.

To show that two graphs are isomorphic, we must provide a function and argue that the function has the necessary properties. To show that two graphs are not isomorphic, it is sufficient to find a *graph-theoretic property* that only one of the graphs has. A graph-theoretic property is a property that is preserved under *permissible* transformations, such as moving the vertices around and making the edges longer or shorter. Some of the simplest graph-theoretic properties are the following:

- How many vertices a graph has.
- How many edges a graph has.
- How many vertices of a certain degree a graph has.

– Whether a graph has three vertices that are all adjacent to each other.

Are the following graphs isomorphic? No, they are not isomorphic. The rightmost graph has fewer vertices than the leftmost graph, and therefore no function from the leftmost graph to the rightmost graph can be *injective*, or *one-to-one*:

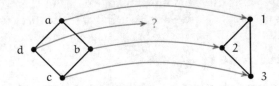

Are the following graphs isomorphic? No, they are not isomorphic. The rightmost graph has more vertices than the leftmost graph, and therefore no function from the leftmost graph to the rightmost graph can be *surjective*, or *onto*:

Are the following graphs, G and H, isomorphic? Yes, they are isomorphic. The function $f : V(G) \rightarrow V(H)$ given by $f(a) = 1, f(b) = 2, f(c) = 3$, and $f(d) = 4$ is an isomorphism. We see that u and v are adjacent in G if and only if $f(u)$ and $f(v)$ are adjacent in H:

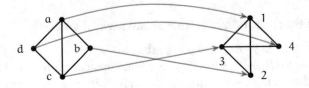

Determining whether two graphs are isomorphic is to find out whether the graphs we are dealing with are essentially the same. If a number of graphs are given, we may want to find out whether any of them are isomorphic to avoid doing excess work. To this day, no one has found an *efficient* method for deciding whether two arbitrary graphs are isomorphic. We know a lot about many special cases, for example trees. In practice, we are able to create quite effective methods, but in the worst case, we have to check all possible functions from one graph to another. Finding isomorphisms from a graph to itself is also a way to uncover *symmetries*.

Exercises

21.1 Find all nonisomorphic graphs with four vertices.

21.2 Check for each of the graphs below that the sum of the degrees of the vertices equals twice the number of edges:

21.3 Suppose that a graph has 10 vertices and that each vertex has degree 6. How many edges does the graph have?

21.4 Can there be a graph with 5 vertices such that each vertex has degree 3?

21.5 Draw a graph whose vertices have the following degrees, or explain why there is no such graph.

(a) 5, 4, 3, 3, 2

(b) 3, 3, 3, 3, 2

(c) 3, 2, 2, 1, 0

(d) 4, 4, 3, 2, 1

(e) 5, 4, 3, 2, 1

(f) 1, 1, 1, 1, 1

21.6 If a simple graph G has n vertices and m edges, how many edges does $\overline{G}$ have?

21.7 Prove that there are $\binom{n}{2}$ edges in a complete graph with n vertices.

21.8 Prove that there are $n(n-1)/2$ edges in a complete graph with n vertices using the fact that the sum of the degrees of the vertices equals twice the number of edges.

21.9 Prove that there is always an even number of vertices of odd degree in a given graph. Hint: It follows from the first result about the degrees of the vertices of a graph.

21.10 Provide an argument that the following graphs are not isomorphic. Hint: Find a property that only one graph has.

21.11 Decide whether the following pairs of graphs are isomorphic:

   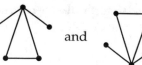

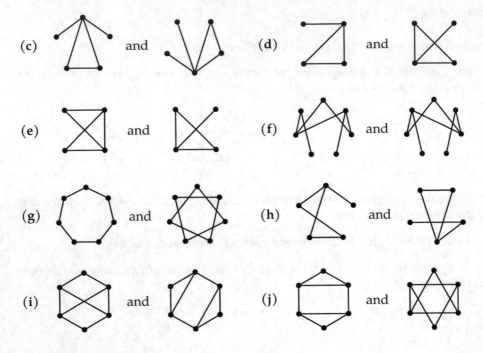

21.12 We defined isomorphism only for *simple* graphs. How can we define isomorphism for graphs that are not simple?

21.13 Prove that if G and H are isomorphic graphs, then $\overline{G}$ and $\overline{H}$ are also isomorphic.

21.14 Find a self-complementary graph with eight vertices.

21.15 A graph is **planar** if it is possible to draw it in two dimensions in such a way that no edges intersect each other. For each of the following graphs, find an isomorphic graph that is planar:

21.16 Explain why graphs can be axiomatized by means the formulas $\forall x \neg Rxx$ and $\forall x \forall y (Rxy \rightarrow Ryx)$.

21.17 We say that a graph is n-**regular** if the degree of each of the vertices is n. Give examples of n-regular graphs for $n = 1, 2, \ldots, 5$.

Chapter 22

Walks in Graphs

■■·■■■■■■■■■■■■■■■■■■■■)■■

In this chapter you will learn more about graph theory, especially about different ways to walk around in graphs. We will define walks, paths, trails, circuits, cycles, and trees. Through these concepts you will learn more about basic properties of graphs.

The Bridges of Königsberg

Our next topic deals with structural properties of graphs and different ways of walking around in graphs. We will start with the classic example of the *bridges of Königsberg* (now Kaliningrad, in Russia). In short, we have seven bridges connecting four areas of land. The question is whether it is possible to take a walk in this city in such a way that one crosses each of the seven bridges *exactly* once. This problem is known to have been solved by the Swiss mathematician and physicist *Leonhard Euler* (1707–1783) around 1735. He did not himself use our modern graph-theoretic terminology, but we shall see that the problem is the same as finding an *Eulerian trail* in the multigraph representing Königsberg. We can represent the situation in the following way:

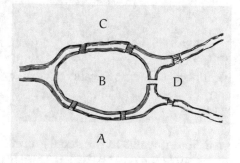
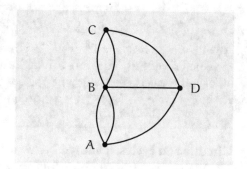

This is a multigraph because we have parallel edges between A and B, and B and C. The question is now, *Is it possible to walk over all the edges in this graph exactly once?* We first look at some of the ways to make what it means to "walk over the edges in a graph" more precise.

© Springer Nature Switzerland AG 2021
R. Antonsen, *Logical Methods*, https://doi.org/10.1007/978-3-030-63777-4_23

Paths and Circuits

Definition 22.1. Walks and paths

A **walk** of length n in a graph is a sequence of vertices and edges of the form

$$v_0 e_1 v_1 e_2 v_2 \ldots e_n v_n,$$

where e_i is an edge that connects v_{i-1} and v_i for $i \in \{1, 2, \ldots, n\}$. We say that the sequence goes *from* v_0 *to* v_n. If no vertex occurs more than once, it is called a **path**.

When a graph is simple, there are no parallel edges, and we can then denote walks as sequences of vertices, $v_0 v_1 v_2 \ldots v_n$. Notice that the *length* of a walk is the same as the number of edges in the walk. A single vertex is a walk of length zero. The terminology of graph theory varies with the context; the word "trail" is sometimes used for a walk in which no edge occurs more than once.

Q **Walks and paths.** Pictured below are two different walks from 1 to 4 in the same graph. The walk in the left graph, 125634, is a path, but the walk in the right graph, 1256325634, is not a path, because it goes through the vertices 2563 twice:

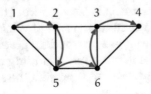 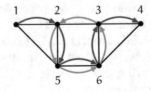

Definition 22.2. Circuits and cycles

A walk whose first and last vertices coincide is called a **closed** walk. A closed walk in which no edge occurs more than once is called a **circuit**. A path $v_1 v_2 \ldots v_n$ with at least three vertices along with the edge that connects v_n and v_1 is called a **closed path** or a **cycle** of length n. A graph without any cycles is called **acyclic**.

Q **Circuits and cycles.** Pictured below are three different walks in the same graph. The walk 254123652 is closed, but not a circuit, because the edge between 2 and 5 occurs twice. The walk 2412362 is a circuit, because it is closed and no edge occurs more than once. It is not a cycle, because the vertex 2 occurs twice. The walk 26542 is a cycle, because it is a path with at least three vertices, 2654, along with the last edge from 4 to 2. The walks 12541, 1241, 23652, and 2362 are some of the other cycles in this graph:

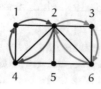

Closed walk   *Circuit*   *Cycle*

It is common to identify circuits and cycles that differ only with respect to their starting node or order. For example, there are eight different ways to describe the cycle 12541, depending on where we start and in what order we go through the vertices: 12541, 25412, 54125, 41254, 14521, 45214, 52145, and 21452. All of these represent the same cycle. ♦

Definition 22.3. Connected

A graph is **connected** if there is a path between all pairs of vertices in the graph, that is, if it is possible to get from each vertex to any other vertex by following the edges.

Two connected graphs.   *Two disconnected graphs.*

When we study graphs, it is usually sufficient to look at connected graphs, because each disconnected graph can be divided into connected **components**. Using this concept we can also define *trees*, which are a special type of graph.

Definition 22.4. Tree

A **tree** is a connected acyclic graph. A vertex of degree one in a tree is called a **leaf node**. A graph all of whose components are trees is called a **forest**.

Two trees.   *Two graphs that are not trees.*

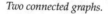

Trees have many interesting properties. Here are some of them:

– If a tree has n vertices, it has n − 1 edges.
– Every tree with at least two vertices has at least two leaf nodes.
– If you remove an edge from a tree, the graph is no longer connected.
– Between any two vertices in a tree there is exactly one path.

Eulerian Trails and Circuits

Definition 22.5. Eulerian trail/circuit

Let G be a connected graph. An **Eulerian trail** is a walk that contains each edge of G exactly once. An **Eulerian circuit** is an Eulerian trail whose first and last vertices coincide. A connected graph that has an Eulerian circuit is called an **Eulerian** graph.

Note. An *Eulerian trail* is also often called an *Eulerian path*, but we have defined a *path* as a walk in which no vertex occurs more than once. In an Eulerian trail we are allowed to repeat vertices, but not edges.

Q   **Eulerian trails and circuits.** Here are three different graphs:

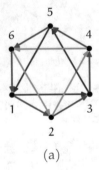 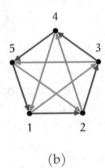 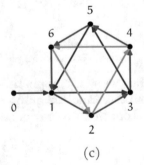

(a)   (b)   (c)

In graph (a) we find the Eulerian circuit 1351234562461, and in graph (b) we find the Eulerian circuit 12345135241. We notice that all the vertices in these graphs have degree four, which is an even number. If we add an extra vertex to (a), we get the graph in (c). *Is there an Eulerian trail or an Eulerian circuit in this one?* We find more Eulerian trails, for example 01351234562461, and we see that every Eulerian trail must begin or end at vertex 0, because if you arrive there, there is no remaining vertex by which to depart. We cannot find any Eulerian circuit, and it is because of vertex 0. We observe that this vertex has degree one, which is an odd number. ♦

Q   **The bridges of Königsberg.** The bridges of Königsberg problem now becomes, *Does the multigraph representing Königsberg have an Eulerian trail?*

An Eulerian trail must go through the vertex A at least once, for example DAB. If we go into the vertex A and out again, only one edge remains, for example BA. In order to go over all three edges that are incident to A, we must begin or end at A. The same applies to B, C, and D, because they all have *odd degree*. We can conclude that the graph does not have an Eulerian trail.   ♦

In the examples above it appears that there is a connection between the existence of Eulerian trails/circuits and whether the degrees of the vertices are even numbers. Therefore, suppose we have a graph with an Eulerian circuit. *What, then, are the degrees of the vertices in the graph?* We can write an Eulerian circuit as a sequence $v_0 e_1 v_1 e_2 v_2 \ldots e_n v_n$, where $v_n = v_0$, and this gives us directly, by looking at the sequence, that the degree of a vertex must be twice the number of times it occurs in the sequence. This means that each vertex in the graph must have degree equal to an even number.

But the reverse is also true, given that G is a connected graph:

- If the degree of each vertex of G is an even number, then G contains an Eulerian circuit.
- If exactly two vertices in G have odd degree, then G contains an Eulerian trail from each of those vertices to the other.
- If G has more than two vertices of odd degree, the graph does *not* contain an Eulerian trail.

Suppose that G is a connected graph in which the degree of each vertex is an even number. We can now construct an Eulerian circuit in the following way: Select any vertex u and create a walk by following the edges around the graph. Sooner or later, we must come back to u, and we have thus constructed a closed walk. The reason we have to come back to u is that all the vertices have even degree. Each time we come across a vertex in the walk that is not u, there is an unused edge along which we can proceed, and sooner or later we must return to u. If this walk is not an Eulerian circuit, there must be a vertex v in the walk that is adjacent to a vertex outside of the walk. By creating a new walk from v with only unused edges and then combining the two the walks together into one, and continuing in this way as necessary, we finally obtain an Eulerian circuit. We can argue in the same way for the other two assertions.

Digression

The **four-color problem** is a mathematical problem with an interesting history. The problem is about how many colors you need to color a map without two adjacent regions having the same color. If we represent the regions as vertices and the borders as edges, this is in fact a graph-theoretic problem. Let us examine the following maps:

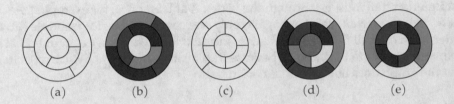

(a)   (b)   (c)   (d)   (e)

Map (a) is an example of a map that requires four colors, and (b) shows one way to color the map such that no adjacent regions have the same color. The question is whether maps exist that require five colors. The map in (c) can be colored with five colors, as in (d), but in (e) we see that four colors suffice.

The four-color problem was first formulated in 1852 by the South-African mathematician *Francis Guthrie* (1831–1899). In 1879, the British mathematician *Alfred Kempe* (1849–1922) wrote a "proof" that four colors are sufficient, but 11 years later a mistake was found, and the problem remained open until 1976, when the American mathematician *Kenneth Appel* (1932–2013) and German mathematician *Wolfgang Haken* (1928–) finally found a proof. They reduced the problem to checking a specific property for 1936 different maps, and they used a computer program to check each of those maps. This is the first example in history of a mathematical proof carried out in part by a computer program. Because the computer program was large, it was difficult to check whether the proof was correct. Many mathematicians therefore believed that theirs was not a proper *proof*. But no one found anything wrong with the proof, and it was generally accepted that the four-color problem had actually been solved. In 1995, a simpler proof was found, but it was still based on a computer program. In 2004, there was a breakthrough when the Canadian mathematician *Georges Gonthier* (1962–) gave a *formal proof* with the interactive theorem prover *Coq*. The formalization removed all doubt surrounding the problem and indirectly gave rise to novel mathematics. *Coq* is an example of software designed to help with finding and checking mathematical proofs, testing hypotheses, finding counterexamples, and ensuring that proofs are correct. Many such *interactive theorem provers* are programmed with relatively little code, so that the program itself can be checked.

Hamiltonian Paths and Cycles

We move on to paths containing *all* the vertices in a graph, regardless of whether all the edges are included. Such paths are called **Hamiltonian paths** and are named after the Irish mathematician and physicist *William Rowan Hamilton* (1805–1865). Finding Hamiltonian paths is much more difficult than finding Eulerian trails.

> **Definition 22.6. Hamiltonian path / cycle**
>
> Let G be a connected graph. A **Hamiltonian path** is a path containing each vertex of G exactly once. A **Hamiltonian cycle** is a cycle containing each vertex of G exactly once. A graph with a Hamiltonian cycle is called **Hamiltonian**.

Q **The traveling salesman problem.** Suppose a set of cities is given and that all the distances between pairs of cities are known. The **traveling salesman problem** is then, *What is the shortest route that visits every city exactly once and begins and ends in the same city?* This is the same as finding the "shortest" Hamiltonian cycle in a weighted complete graph, where "weighted" means that each edge has a value. Ever since the 1930s, this problem has been discussed and analyzed in mathematics, and it is a difficult problem. The reason it is so difficult is that the number of potential Hamiltonian cycles grows rapidly. If a graph has n vertices, there are $n!$ permutations of these vertices, and the "brute-force" solution of checking every such permutation quickly becomes impractical. If we managed to check one billion permutations every second, it would take around 77 years to check 20! permutations. If we wanted to check 28! permutations at this speed, it would take time equal to the current age of the universe, which is about 14 billion years. And that is for only 28 vertices. ◆

Q **Hamilton's puzzle.** This is a question based on the graph that represents a dodecahedron, one of the five Platonic solids, where each vertex represent a corner of the dodecahedron and each edge represents an edge: *Is there a Hamiltonian cycle in this graph?* The answer is yes, and we can see this in the following illustration.

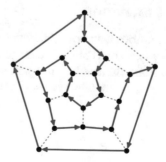

Q   **The knight's tour problem.**   *Is it possible to move a knight on a chessboard such that the knight visits each square exactly once?* Such a sequence of moves is called a "knight's tour." Let the **knight's graph** for 8×8 squares be the left graph below, where each vertex represents a square on a chessboard and each edge represents a legal knight's move.

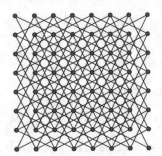 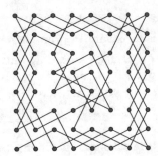

Visiting each square on a chessboard exactly once is the same as finding a Hamiltonian path or cycle in the graph. One of the first mathematicians to study this problem was *Leonhard Euler*, who in 1759 published an article on the knight's tour problem with the title "Solution of a curious question which does not seem to have been subjected to any analysis." The first Hamiltonian path in this article is illustrated in the right graph above. It was not until 2014 that someone was able to calculate the exact number of Hamiltonian paths in the graph, which is 19 591 828 170 979 904.   ♦

No one has managed to construct an *efficient* algorithm to determine whether there is a Hamiltonian cycle in a graph. This is "as difficult" as deciding whether a propositional formula is a tautology. Both of these belong to the class of **NP**-complete problems. In practice, it is rare that we really need to find a Hamiltonian cycle. There are also many special cases and heuristics we can take advantage of. Often it is sufficient to find an Eulerian circuit, or the problem can be made easier if one is allowed to traverse some of the vertices several times.

Final Comments

This concludes our brief journey through graph theory. These two chapters constitute only a small taste of a large and interesting field. Some topics we have not talked about are *weighted* graphs, *infinite* graphs, *coloring* of graphs, *random walks*, trees that *span* graphs, *subgraphs*, the analysis of *planar* graphs, and much, much more.

Exercises

22.1 Draw all *simple* and *connected* graphs with four vertices. Hint: There is an even number of them.

22.2 (a) Do the following graphs have *Eulerian circuits* or *Eulerian trails*? If yes, provide at least one example; if no, explain why such cannot be found.

(b) Do the graphs have *Hamiltonian paths* or *Hamiltonian cycles*? If yes, provide at least one example.

  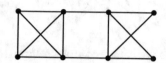 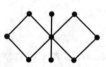

22.3 How many different Eulerian circuits can you find in this graph?

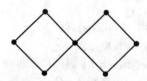

22.4 Find a *simple* graph that represents the problem with the bridges of Königsberg.

22.5 (a) Draw all six trees with six vertices.

(b) Draw all eleven trees with seven vertices.

(c) Can you get a tree by *adding* an edge to a tree?

(d) Can you get a tree by *removing* an edge from a tree?

(e) Prove that a graph G is a tree if and only if there is exactly one path between each pair of vertices of G.

22.6 (a) Provide an inductive definition of the set of trees.

(b) What is the difference between the trees in graph theory and the binary trees we encountered in Chapter 9 (*page 104*)?

22.7 Let ~ be defined on the set of vertices in a graph G such that u ~ v holds if there is a walk from u to v. Show that ~ is an equivalence relation. What are the equivalence classes of ~?

22.8 (a) Prove that if G is not connected, then $\overline{G}$ must be.

(b) Find a proof or a counterexample for the following statement: For every graph G, either G or $\overline{G}$ is a Hamiltonian graph.

22.9 Prove that in a group of six people, there will always be three mutual acquaintances or three mutual strangers.

22.10 A graph is called **bipartite** if the set of vertices can be divided into two disjoint subsets U and V such that each edge connects a vertex of U with a vertex of V. The following are some examples of bipartite graphs.

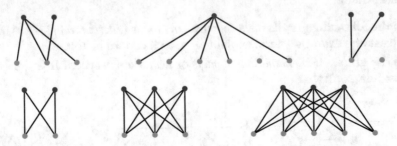

The bottom three graphs are called **complete bipartite graphs**, because each vertex of one set is adjacent to each vertex of the other set. If one set has m vertices and the other set has n vertices, the complete bipartite graph is denoted by $K_{m,n}$. Shown here are $K_{2,2}$, $K_{3,3}$, and $K_{3,5}$.

(a) How many vertices are there in $K_{m,n}$?

(b) How many edges are there in $K_{m,n}$?

(c) Are there any *Eulerian circuits* or *Eulerian trails* in $K_{2,3}$?

(d) Prove that a graph is bipartite if and only if each cycle in the graph has length equal to an even number.

22.11 The so-called **Petersen graph** has many special properties. Here it is drawn in two different ways:

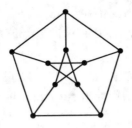
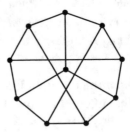

(a) Show that the above graphs are isomorphic.

(b) Find cycles of lengths 6, 8, and 9.

(c) Are there any *Eulerian circuits* or *Eulerian trails* in the graph?

(d) Are there any *Hamiltonian paths* or *Hamiltonian cycles* in the graph?

Chapter 23

Formal Languages and Grammars

■■■■■■■■■■■■■■■■■■■■■■■■■■

In this chapter you will learn more about formal languages. You will learn about the relationship between regular languages and regular expressions, how deterministic finite automata can be used to characterize regular languages, and finally a bit about regular and context-free grammars.

Formal Language Theory

The study of formal languages and grammars has its roots in linguistics and logic, but it is also an important part of both computer science and mathematics. In linguistics, formal languages are used, among other things, to analyze syntax and sentence structures in natural languages. In computer science, formal languages are used to translate from programming languages to machine code, and to prove properties of programs and systems. In more theoretical computer science, formal languages are used to study the complexity of algorithms and calculations. Although formal language theory is only about sets of strings over alphabets, there are numerous areas of application.

In Chapter 9 *(page 105)* we looked at how a language could be inductively defined as a set of strings over an underlying alphabet. Then there were no restrictions, provided that we defined a language precisely. Now we will look at some practical and widely used ways to describe languages. Many of these give rise to effective methods for determining whether a particular string is in a language. We will see that some languages are simple to describe, while others are not.

Operations on Languages

If two languages L and M are given, we can take the union $L \cup M$ of them. This is one of the simplest operations on languages. We will now look at two other ways to construct languages: taking the *concatenation* and the *closure*.

We have already talked a lot about *concatenation* as an operation on *strings*. For example, the concatenation of aaa and bbb is aaabbb, and for all strings s and the empty string Λ, we have that $\Lambda s = s = s\Lambda$. The notation s^n denotes the string s concatenated with itself n times. For example, $(ab)^3$ stands for ababab, and a^3b^3 stands for aaabbb. We will now define the corresponding operation on *languages*:

Definition 23.1. Concatenation of two languages

If L and M are languages, the **concatenation** of L and M is the language $LM = \{st \mid s \in L \text{ and } t \in M\}$.

Q **The concatenation of two languages.** Let $L = \{aa, b\}$ and $M = \{c, dd, e\}$. The concatenation of L and M is the language $LM = \{aac, aadd, aae, bc, bdd, be\}$. This is different from the language $ML = \{caa, cb, ddaa, ddb, eaa, eb\}$. Concatenation is therefore not a commutative operation. ◆

For example, we have that $L\{\wedge\} = \{\wedge\}L = L$ for all languages L, which means that $\{\wedge\}$ is the identity for concatenation. Concatenating with an empty language always gives an empty language: for all L, $L\varnothing = \varnothing L = \varnothing$. Concatenation is *associative* because the placement of parentheses does not matter: It is always the case that $(LM)N = L(MN)$. For example, $(\{a\}\{b\})\{c\}$ equals $\{a\}(\{b\}\{c\})$. Both languages are $\{abc\}$. We can therefore drop the parentheses and write LMN.

Taking the concatenation of a language L with itself, gives the language $LL = \{st \mid s \in L \text{ and } t \in L\}$, and we can define L^n recursively by saying that $L^0 = \{\wedge\}$ and $L^{n+1} = LL^n$. For all positive numbers n, we get that L^n equals the set $\{s_1 s_2 \ldots s_n\}$, where each s_i is in L. The union of all these is called the *closure* of L.

Definition 23.2. Closure of a language

If L is a language, the **Kleene closure** – or just the **closure** – of L is the language $L^* = L^0 \cup L^1 \cup L^2 \cup L^3 \cup \cdots$, that is, the set of all finite strings over L.

The closure operation, which is also called **Kleene star**, is named after the American mathematician *Stephen Cole Kleene* (1909–1994), who was one of the first to study such formal languages.

Q **The closure of a language.** Let L be the language $\{1, 22\}$. Then we get that $L^0 = \{\wedge\}$. Using the definition of L^n, we get

$$L^1 = LL^0 = \{1, 22\}\{\wedge\} = \{1, 22\},$$
$$L^2 = LL^1 = \{1, 22\}\{1, 22\} = \{11, 122, 221, 2222\},$$
$$L^3 = LL^2 = \{1, 22\}\{11, 122, 221, 2222\}$$
$$= \{111, 1122, 1221, 12222, 2211, 22122, 22221, 222222\}.$$

The language L^* consists of the union of all the L^n. ◆

We thus have three important operations on languages: we can take the *union* of languages, the *concatenation* of languages, and the *closure* of languages. By restricting ourselves to these three, we get a widely used type of language, the *regular languages*.

Regular Languages

Definition 23.3. Regular language

The set of **regular languages** over an alphabet A is the smallest set such that the following holds:

- ∅ and {Λ} are **regular languages** over A, and {a} is a **regular language** over A for all a ∈ A.
- If L and M are regular languages, then L ∪ M, LM, and L* are also **regular languages**.

Regular languages are inductively defined. Notice that the base set consists of the empty language, the language consisting of the empty string, and one language for each character of the alphabet.

Q **Regular languages.** Let the alphabet A be the set {a, b}. Let us try to find the regular languages over A. From the base case of the definition, we get that the following are four different regular languages:

$$\varnothing, \quad \{\Lambda\}, \quad \{a\}, \quad \{b\}.$$

The inductive step gives us several ways to create new languages.

By taking the union of languages from the base case, we get the following languages:

$$\{a, b\}, \quad \{\Lambda, a\}, \quad \{\Lambda, b\}, \quad \{\Lambda, a, b\}.$$

By concatenating languages, we get, for example,

$$\{b\}\{\Lambda, a\} = \{b, ba\} \quad \text{and} \quad \{b\}\{a, b\} = \{ba, bb\}.$$

By taking the closure of languages, we get, for example,

$$\{b\}^* = \{\Lambda, b, bb, bbb, \dots\},$$
$$\{a, bb\}^* = \{\Lambda, a, bb, abb, bba, aa, bbbb, \dots\}.$$

We can now ask whether $\{aa, aba, abba, \dots, ab^na, \dots\}$ is a regular language. On reflection, we see that the language equals the concatenation $\{a\}\{b\}^*\{a\}$. So it is a regular language. In the same way, the language that consists of all strings of the form a^nb^m is regular, because it is equal to the concatenation $\{a\}^*\{b\}^*$. ◆

This example shows that there are many different regular languages over a given alphabet. Now we will look at a practical and useful notation that can be used to describe regular languages.

Regular Expressions

Definition 23.4. Regular expression

The set of **regular expressions** over an alphabet A is the smallest set such that:

- $\emptyset$ and $\wedge$ are **regular expressions** over A, and a is a **regular expression** over A for all $a \in A$.
- If R and S are regular expressions, then (R), $R|S$, RS, and R^* are also **regular expressions**.

This is also an inductively defined set, and it is similar to the definition of regular languages, but here it is syntactic *expressions* that are defined. We think of regular expressions as certain strings over the alphabet $A \cup \{|, ^*, \wedge, \emptyset, (,) \}$. The relationship between regular expressions and regular languages is one we have seen before: regular expressions are syntactic and will be *interpreted* as regular languages. We will now see that each regular expression describes a regular language.

Q Regular expressions. Here are some simple regular expressions over the alphabet $\{0, 1\}$:

$$\wedge, \quad \emptyset, \quad 1, \quad 0, \quad \wedge|1, \quad 1^*, \quad 1|01, \quad (1|0)1. \qquad \blacklozenge$$

Assumption. In order to avoid too many parentheses, we use the following **precedence rules**, or **order of precedence**, for regular expressions: We assume that * binds most strongly, that concatenation binds more weakly than *, and that | binds the most weakly. We therefore read 01^* as $0(1)^*$ and not $(01)^*$, and $0|12$ as $0|(12)$ and not $(0|1)2$.

Interpretation of Regular Expressions

A regular expression describes and gives rise to a regular language in the following way.

Definition 23.5. Interpretation of regular expressions

Let A be an alphabet. We recursively define a function L from the set of regular expressions over A to the set of regular languages over A in the following way:

- $L(\emptyset) = \varnothing$
- $L(\wedge) = \{\wedge\}$
- $L(a) = \{a\}$, for each $a \in A$

- $L(R|S) = L(R) \cup L(S)$ (*union*)
- $L(RS) = L(R)L(S)$ (*concatenation*)
- $L(R^*) = L(R)^*$ (*closure*)

In addition, the parentheses have no other function than grouping: $L((R)) = L(R)$. For each regular expression E, we get a regular language $L(E)$ in this way.

Notice that the structure of the definition of the function L is exactly the same as in the inductive definition of regular languages. Also notice how a regular expression R^* is interpreted: the last item in the definition says that $L(R^*) = L(R)^*$, and this means that we first find $L(R)$, which is a regular language, and that we then take the closure of this language.

Q **Interpretation of a regular expression.** Look at the regular expression $a\,|\,b^*c$. We find the language that this expression defines in the following way:

$$
\begin{aligned}
L(a\,|\,b^*c) &= L(a) \cup L(b^*c)\\
&= L(a) \cup L(b^*)L(c)\\
&= L(a) \cup L(b)^*L(c)\\
&= \{a\} \cup \{b\}^*\{c\}\\
&= \{a\} \cup \{\Lambda, b, bb, bbb, \dots\}\{c\}\\
&= \{a\} \cup \{c, bc, bbc, bbbc, \dots\}\\
&= \{a, c, bc, bbc, bbbc, \dots\}.
\end{aligned}
$$
♦

Q **Regular expressions in practice.** Many programming languages and programs are created with support for using regular expression to search text. You will see that the notations and standards for regular expressions vary, but the theory is the same. In the POSIX standard, for example, the brackets [and] are used to express union: here [hcr]at stands for {hat, cat, rat}. There is also a number of special symbols that represent the beginning ^ and the end $ of a line, which makes text search easier.
♦

Q **Interpretation of several regular expressions.** Here are more examples of how regular expressions are interpreted:

- 0^*1^* describes the language that consists of all strings over $\{0, 1\}$ in which all occurrences of 0 are to the left of all occurrences of 1.
- $(01)^*$ describes the language $\{\Lambda, 01, 0101, 010101, \dots\}$.
- $(0\,|\,1)^*$ describes the language of *all* strings over $\{0, 1\}$.
- $0\,|\,1^*$ describes the language $\{0, \Lambda, 1, 11, 111, \dots\}$.
- $(00)^*$ describes the language $\{\Lambda, 00, 0000, 000000, \dots\}$.
- 00^* describes the language $\{0, 00, 000, \dots\}$.
  This is often written as 0^+, with + instead of *.
- $0(\Lambda\,|\,1)0$ describes the language $\{00, 010\}$.
  That 1 is optional is often written as $1^?$, with ? instead of *.

We can also interpret the regular expressions from the previous example:

$$
\begin{aligned}
L(\Lambda) &= \{\Lambda\}, & L(\emptyset) &= \emptyset,\\
L(1) &= \{1\}, & L(0) &= \{0\},\\
L(\Lambda\,|\,1) &= \{\Lambda, 1\}, & L(1^*) &= \{\Lambda, 1, 11, \dots\},\\
L(1\,|\,01) &= \{1, 01\}, & L((1\,|\,0)1) &= \{11, 01\}.
\end{aligned}
$$
♦

Deterministic Automata

The British mathematician and logician *Alan Turing* (1912–1954) studied, among other things, the basic questions *What is an algorithm?* and *What is computable?* In 1936 he invented what today we call a **Turing machine**, which is an abstract mathematical model making precise what it means to perform a computation.

Here we will look at a design that is much more limited but that uses many of the same concepts as a Turing machine. We will look at *finite automata*, which can be used for *deciding* whether a string is in a given regular language. This is the beginning of a rich and wonderful theory, of which you will get here only a little taste, along with some examples.

Definition 23.6. Deterministic finite automaton

A **deterministic finite automaton (DFA)** consists of the following:

- A finite set of **states**.
- A finite set of **input symbols**, also called an **alphabet**.
- A **start state** that is one of the states.
- A set of **accepting states**, also called **final states**, that is a subset of the states.
- A **transition function** that takes two arguments – a state and an input symbol – and returns a state.

Such a finite automaton is called *deterministic* because the transition function uniquely determines what the *next state* is when a state and an input symbol are given. We usually draw automata as directed graphs in which the vertices represent the states and the edges represent the transition function.

Q   **Deterministic finite automaton.** Here q_0, q_1, and q_2 are the states. We have one start state, q_0, marked with a "start" arrow, and one accepting state, q_2, marked with a green color and a double edge:

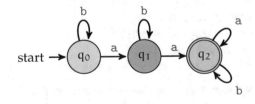

We say that the string abbba is *accepted* by the automaton, because we (deterministically) get from the initial state to an accepting state by following the arrows that the characters of the string abbba suggest. We say that the automaton *accepts* strings that lead to accepting states. The string abb is not accepted, because by following the arrows we end up in the state q_1, which is not an accepting state. This automaton accepts exactly the strings over {a, b} that have at least two a's in them. Notice that this language is regular and that it is possible to describe it with the regular expression b*ab*a(a|b)*.   ◆

The example above shows how a deterministic finite automaton will either *accept* a string over the alphabet or not.

Q   **Deterministic finite automaton.** Here is another automaton:

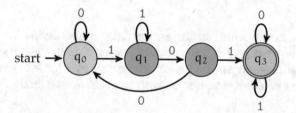

If we start with the string 1101 and follow the edges as we read each character, we end up in q_3. Because this is an accepting state, 1101 is accepted. The string 100 is not accepted, because we end up in state q_0, which is not an accepting state. *Try to find out for yourself which strings are accepted. Can you find a regular expression that describes the same language?*   ◆

Automata and Regular Languages

There is a beautiful connection between regular languages and deterministic finite automata: for each regular language there is a deterministic finite automaton that accepts all the strings in that language and only those strings. The opposite is also the case: if we have a deterministic finite automaton, the strings that the machine accepts make up a regular language. This means that we have characterized the regular languages in two different ways, via both regular expressions and deterministic finite automata.

Nondeterministic Automata

There are also *nondeterministic* finite automata, and for these there is no requirement that there be a transition for each character of the alphabet. It is also permitted to have a transition associated with the empty string. The precise definition is similar to that for deterministic finite automata, but the transition function returns a *set* of states instead of one state. This means that the state we are in is ambiguous as we read an input string.

Q **Nondeterministic finite automaton.** The following automaton accepts only strings ending with 10:

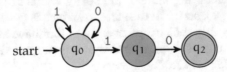

If we read the string 111, we end up either in state q_0 or in state q_1, and because neither of these is an accepting state, the string is not accepted. If we read the string 1110, we are in state q_0 or q_2, and because q_2 is an accepting state, the string is accepted. This language is also described by the regular expression (0|1)*10. ◆

Even though nondeterministic finite automata can be defined much more freely than the deterministic ones, it is actually impossible to describe anything more than the regular languages. It is possible to prove that for every nondeterministic automaton there is a deterministic automaton that does exactly the same job.

Formal Grammars

The last topic in formal language theory is *formal grammars*. Briefly, a formal grammar is a set of rules that describe a language by saying how strings can be *rewritten* into other strings. For example, the grammar can say that the symbol S must be rewritten to 00A, and that A must be rewritten to 11. We then get 0011 from S.

Definition 23.7. Formal grammar

A **formal grammar**, or a **grammar**, consists of the following:

  - A finite set of **terminal symbols**.
  - A finite set of **nonterminal symbols**.
  - A **start symbol**, one the nonterminal symbols.
  - A finite set of **production rules** of the form $X \rightarrow Y$, where X and Y are strings of terminal and nonterminal symbols, and X cannot be the empty string.

We assume that the sets of terminal and nonterminal symbols are disjoint and that there is at least one production rule with only the start symbol on the left side.

Think of the terminal symbols as a set of *constants* or an *alphabet*, for example {a, b}, and the nonterminal symbols as *variables* or *placeholders*. It is common to use lowercase letters for terminal symbols and uppercase letters for nonterminal symbols. We first look at a simple example:

Q **A grammar for a regular language.** Let $\{a, b\}$ be the set of terminal symbols, S the only nonterminal symbol, and let the production rules be $S \rightarrow \wedge$ and $S \rightarrow abS$. We shorten this to $S \rightarrow \wedge \mid abS$. This grammar describes a language over $\{a, b\}$ that we find by starting with the start symbol S, and *deriving* the strings in the language:

- $S \Rightarrow \wedge$ is a derivation of $\wedge$.
- $S \Rightarrow abS \Rightarrow ab$ is a derivation of ab.
- $S \Rightarrow abS \Rightarrow ababS \Rightarrow abab$ is a derivation of $abab$.

For each *part* of the derivation, exactly one of the production rules is used. For example, in $abS \Rightarrow ababS$, the production rule $S \rightarrow abS$, which says that S can be rewritten to abS, is used. The language defined by this grammar is the set of all strings of the form $(ab)^n$. This is a regular language that is also described by the regular expression $(ab)^*$. ◆

Definition 23.8. Derivations

Suppose that a grammar is given and that M is the set of all strings over terminal and nonterminal symbols of the grammar. Let the binary relation $\Rightarrow$ be defined on M such that $AXB \Rightarrow AYB$ holds exactly when $X \rightarrow Y$ is a production rule of the grammar, and A and B are strings in M. A sequence of the form $x_1 \Rightarrow x_2 \Rightarrow \cdots \Rightarrow x_n$ is called a **derivation** of x_n from x_1, and we say that x_n can be derived from x_1. The **language** that is defined by the grammar is the set of strings that can be derived from the start symbol S and contain only terminal symbols.

Notice that the production rules of a grammar already form a relation on the set of all strings, but that the relation is *extended* to $\Rightarrow$ by allowing that *substrings* also can be rewritten. Different restrictions on grammars define different types of languages. Finally, in this chapter, we will look at *regular* and *context-free* grammars.

Definition 23.9. Regular grammar

A **regular grammar** is a grammar in which each production rule is of the form $A \rightarrow X$, where A is a nonterminal symbol and X contains at most one nonterminal symbol, which must be the rightmost symbol in X.

Q **Regular grammars.** The grammar given by $S \rightarrow \wedge \mid abS$ is regular. It describes the same language as the regular expression $(ab)^*$. The grammar given by $S \rightarrow \wedge \mid aS \mid bS$ describes the set of all strings over $\{a, b\}$. It is also regular and describes the same set as the regular expression $(a \mid b)^*$. No regular grammar can, for example, have $S \rightarrow AB$, which has two nonterminal symbols, or $A \rightarrow aAb$, because the nonterminal symbol A is not the rightmost symbol. ◆

Q   **Another grammar.** The grammar given by the production rules $S \to AB$, $A \to \wedge \mid$ aA, and $B \to \wedge \mid bB$ is *not* regular, but it nevertheless describes a regular language. Here are two derivations:

- $S \Rightarrow AB \Rightarrow aAB \Rightarrow aB \Rightarrow abB \Rightarrow abbB \Rightarrow abb$
- $S \Rightarrow AB \Rightarrow A \Rightarrow aA \Rightarrow aaA \Rightarrow aaaA \Rightarrow aaa$

The grammar describes the set of all strings over $\{a, b\}$ in which all the a's occur before all the b's. A regular expression for this is a^*b^*.   ♦

Definition 23.10. Context-free grammar

A **context-free grammar** is a grammar in which each production rule is of the form $A \to X$, where A is a nonterminal symbol and X is a string of terminal and nonterminal symbols. A language defined by a context-free grammar is called a **context-free language**.

Q   **A context-free grammar.** The grammar given by $S \to \wedge \mid aSb$ is context-free, but not regular. We derive aabb in the following way: $S \Rightarrow aSb \Rightarrow aaSbb \Rightarrow aabb$. The language defined by this grammar consists of all strings of the form $a^n b^n$ and is a context-free language. It is possible to prove that this language is not regular, but we are not going to do so here.   ♦

Digression

We have now seen parts of the so-called **Chomsky hierarchy** over formal grammars. It was described by the American linguist and philosopher *Noam Chomsky* (1928–) for the first time in 1956. *Type 0* grammars are formal grammars, exactly as we defined them, without any restrictions. The languages described by these are the same as those that a Turing machine can describe. *Type 1* grammars are so-called context-sensitive grammars, and *Type 2* and *Type 3* grammars are the context-free and the regular grammars, respectively, that we have seen.

Exercises

23.1 Let $A = \{0, 1\}$ and $B = \{01, 10, 111\}$. Compute the following languages.

(a) AB

(b) BA

(c) $A \cup B$

(d) $B \cup A$

(e) A^2

(f) B^2

(g) B^*

(h) $A(B \cup A)B$

23.2 There are two languages L whose closures L^* are finite. What languages are these?

23.3 Explain why $L\{\wedge\} = \{\wedge\}L = L$ and $L\varnothing = \varnothing L = \varnothing$ for all languages L.

23.4 Let the following grammar be given:

$$S \rightarrow A \mid B \mid AS \mid BS$$
$$A \rightarrow 0 \mid 2 \mid 4 \mid 6 \mid 8$$
$$B \rightarrow 1 \mid 3 \mid 5 \mid 7 \mid 9$$

(a) What language does this grammar define?

(b) Provide a derivation of 0123.

23.5 For each of the following languages, find a grammar that defines the language:

(a) $\{aa, aaaa, aaaaaa, \ldots, a^{2n}, \ldots\}$

(b) $\{\wedge, ab, abab, ababab, \ldots, (ab)^n, \ldots\}$

(c) $\{ab, aab, aaab, \ldots, a^n b, \ldots\}$

(d) $\{ab, aabb, aaabbb, \ldots, a^n b^n, \ldots\}$

(e) $\{bab, baaab, baaaaab, \ldots, ba^{2n+1}b, \ldots\}$

23.6 Do the following pairs of regular expressions describe the same language? If, for example, the pair consists of $0 \mid 1$ and $1 \mid 0$, the answer is yes, because both describe the language $\{0, 1\}$.

(a) $(01)^*$ and $(10)^*$

(b) $(0 \mid 1)^*$ and $(1 \mid 0)^*$

(c) 01^* and $(01)^*$

(d) 11^* and 1^*

(e) $\wedge \mid 0^*$ and 0^*

(f) $1(\wedge \mid 0^*)$ and 10^*

23.7 Do the following pairs of regular expressions describe the same language for all regular expressions E, F, and G? For example, if the pair consists of E and E^*, the answer is no, because, for example, the regular expressions a and a^* describe two *different* languages. If yes, give a brief justification; if no, find a counterexample.

(a) EF and FE

(b) $E \mid F$ and $F \mid E$

(c) $\wedge E$ and E

(d) $\varnothing E$ and E

(e) $\wedge \mid E$ and $\wedge E$

(f) $E(F \mid G)$ and $EF \mid EG$

(g) $(E \mid F)$ and EF

(h) E^* and $E^* E^*$

(i) E^* and $(E^*)^*$

(j) $\wedge \mid E^* E$ and E^*

(k) $(E^* F)^*$ and $(E \mid F)^* F$

(l) $(EF^*)^*$ and $(E \mid F)^* F$

23.8 What languages are described by the following regular expressions? Find natural descriptions of the languages; for example, the regular expression $11(0|1)^*$ describes the set of all strings beginning with 11.

(a) $(10)^*$

(b) $0(01)$

(c) $0|01$

(d) $1^*|(0000)^*$

(e) $00(0|1)^*$

(f) $(0|10)^*1^*$

(g) $(0^*1^*)^*$

(h) $(0^*1^*)^*1111(1|0)^*$

(i) $(00|01|10|11)^*$

23.9 Let L be a language.

(a) Is it necessarily the case that $L^n \subseteq L^{n+1}$?

(b) Is it necessarily the case that $L^n \subseteq L^*$?

23.10 Find regular expressions for the following languages:

(a) The set of strings over $\{3, 4, 5\}$ containing at least one occurrence of 4 and at least one occurrence of 5.

(b) The set of strings over $\{0, 1\}$ such that the second and fourth characters from the left are 0.

(c) The set of strings over $\{0, 1\}$ that contain an occurrence of 1111.

23.11 Find regular expressions for the following languages over $\{0, 1\}$. For each language find a regular expression E such that the described language equals $L(E)$.

(a) All strings beginning with 0 or 111.

(b) All strings in which 0 and 1 alternate. Hint: All strings can both begin and end with 0 and 1.

23.12 For each of the following languages, find a regular expression that describes the language:

(a) $\{a, b, c\}$

(b) $\{aa, ba, ca\}$

(c) $\{aa, ab, ac, ba, bb, bc\}$

(d) $\{ab, abb, abbb, \dots, ab^n, \dots\}$

(e) $\{a, b, ab, ba, abb, baa, \dots, ab^n, ba^n, \dots\}$

(f) $\{\Lambda, a, abb, abbbb, \dots, ab^{2n}, \dots\}$

23.13 Construct context-free grammars that generate the following languages over $\{0, 1\}^*$:

(a) $\{s \mid s$ begins with 01 and ends with 10 $\}$

(b) $\{s \mid s$ has an odd number of characters $\}$

(c) $\{s \mid s^R = s$, where s^R is the string s reversed$\}$

Chapter 24

Natural Deduction

■■■■■■■■■■■■■■■■■■■■■■■■■■■

In this chapter you will learn about a logical calculus called natural deduction. This calculus is a formalization of logical reasoning and can be used to make derivations and proofs of propositional formulas. You will also learn more about the relationship between syntax and semantics via the concepts of soundness, completeness, and consistency.

Logical Calculi: From Semantics to Syntax

So far, we have mostly talked about semantic properties of formulas, for example that a formula can be valid, satisfiable, contradictory, or falsifiable, and in propositional logic we have used truth tables and mathematical arguments to prove these properties. Another perspective, which is not based on truth tables, is about capturing correct ways of reasoning. Natural deduction is a logical calculus that gives a syntactic characterization of the properties above.

In general, a **logical calculus** consists of inference rules and ways to construct proofs. There are many different logical calculi, and in proof theory we study the properties of these and the differences and similarities between them. Here we will look at a calculus called **natural deduction**. It is called "natural" because it corresponds well with the way we think and reason from assumptions.

Inference Rules of Natural Deduction

Natural deduction consists of a series of inference rules that are used to draw conclusions from assumptions. These are purely syntactic constructions, and they say how **derivations** can be constructed. The set of all derivations is inductively defined based on inference rules. That being said, it is useful to have an interpretation in mind when using the inference rules.

In the beginning, we will limit ourselves to the three connectives $\land$, $\to$, and $\bot$. This is unproblematic, because we can express negation using $\to$ and $\bot$; the formula $F \to \bot$ expresses exactly the same thing as $\neg F$. In natural deduction, and in many logics, $\neg F$ is only an abbreviation for $F \to \bot$, and this is what we will do as well.

© Springer Nature Switzerland AG 2021
R. Antonsen, *Logical Methods*, https://doi.org/10.1007/978-3-030-63777-4_25

An **inference rule**, or just **rule**, in natural deduction typically has the following form:

$$\frac{F \quad F \to G}{G} \to\!\text{E}$$

The formulas above the line are the **premises,** and the formula below the line is the **conclusion**. We say that $\to$ above the line in this case is *eliminated*. The name of the rule appears next to the line. This rule is called $\to$E, which stands for $\to$-*elimination*. We have the following introduction and elimination rules.

The rules for $\wedge$-formulas. We have one introduction rule and two elimination rules for $\wedge$-formulas. We intuitively understand the rules in the following way: if we have a derivation of F and a derivation of G, we also have a derivation of $F \wedge G$, and if we have a derivation of $F \wedge G$, we also have a derivation of F and a derivation of G:

$$\frac{F \quad G}{F \wedge G} \wedge\text{I} \qquad\qquad \frac{F \wedge G}{F} \wedge\text{E} \qquad\qquad \frac{F \wedge G}{G} \wedge\text{E}$$

The rules for $\to$-formulas. We have one introduction rule and one elimination rule for $\to$-formulas. The elimination rule says that if we have a derivation of F and a derivation of $F \to G$, we also have one of G. The introduction rule is different; it says that if we have a derivation of G, where F is an assumption, we also have one of $F \to G$, without the assumption F. This corresponds to our intuition of what $F \to G$ means, that G follows from the assumption F:

$$\frac{F \quad F \to G}{G} \to\!\text{E} \qquad\qquad\qquad \begin{array}{c} [F] \\ \vdots \\ \dfrac{G}{F \to G} \to\!\text{I} \end{array}$$

The rule for $\perp$ and reductio ad absurdum. We have only one rule for $\perp$, and it says that everything can be deduced from $\perp$. If we have a derivation of $\perp$, we also have a derivation of F, no matter what F is. This is a form of elimination rule, because $\perp$ above the line is eliminated. There is no introduction rule for $\perp$, but there is another special rule in which $\perp$ plays a key role, which is the *reductio ad absurdum*-rule or the RAA-rule. This rule is a formalization of proof by contradiction: if it is possible to deduce $\perp$ from $F \to \perp$, then we have a derivation of F without the assumption that $F \to \perp$. This corresponds to assuming $\neg F$ and deriving a contradiction. Then we can conclude with F and remove the assumption $\neg F$:

$$\frac{\perp}{F} \perp \qquad\qquad\qquad \begin{array}{c} [F \to \perp] \\ \vdots \\ \dfrac{\perp}{F} \text{RAA} \end{array}$$

Closing of Assumptions

In two of the rules, $\to$I and RAA, the assumptions are removed, or **closed**, which is indicated by placing brackets around the assumptions. If an assumption is not closed, it is called **open**. To indicate when the various assumptions were closed, we will also place numbers next to the brackets and the names of the rules. For example, look at the $\to$I-rule, the introduction rule for $\to$. If we manage to establish G from an assumption F, we can conclude that $F \to G$.

- *Do we need to keep the assumption F?* No, when we have deduced $F \to G$, that will be the case *without* the assumption F.
- *Is it forbidden to keep the assumption F?* No, but it is unnecessary.

The basic guideline is that we close assumptions as often we can, because we want to obtain a derivation with as few open assumptions as possible.

Q   **Derivation.** The following is a derivation of $P \wedge Q \to Q \wedge P$ in which all the assumptions are closed:

$$\cfrac{\cfrac{[P \wedge Q]^1}{Q} \wedge_E \qquad \cfrac{[P \wedge Q]^1}{P} \wedge_E}{\cfrac{Q \wedge P}{P \wedge Q \to Q \wedge P} \to_{I_1}} \wedge_I$$

This derivation is constructed *step by step* based on the formula $P \wedge Q$. Using the $\wedge_E$-rule twice, on two occurrences of $P \wedge Q$, we get two different derivations:

$$\cfrac{P \wedge Q}{Q} \wedge_E \qquad\qquad\qquad\qquad \cfrac{P \wedge Q}{P} \wedge_E$$

The $\wedge_I$-rule now says that we can merge these into a single derivation in the following way:

$$\cfrac{\cfrac{P \wedge Q}{Q} \wedge_E \qquad \cfrac{P \wedge Q}{P} \wedge_E}{Q \wedge P} \wedge_I$$

This derivation has two open assumptions, both of which equal $P \wedge Q$. Using the $\to$I-rule, we can now create a new derivation, with $P \wedge Q \to Q \wedge P$ as conclusion and in which all occurrences of $P \wedge Q$ are closed:

$$\cfrac{\cfrac{[P \wedge Q]^1}{Q} \wedge_E \qquad \cfrac{[P \wedge Q]^1}{P} \wedge_E}{\cfrac{Q \wedge P}{P \wedge Q \to Q \wedge P} \to_{I_1}} \wedge_I$$

♦

Q **Derivation.** The following shows how a derivation of $P \to ((P \to \bot) \to \bot)$, or $P \to \neg\neg P$, can be constructed from the formulas P and $P \to \bot$:

$$\frac{P \qquad P \to \bot}{\bot} \to_E \qquad\qquad \frac{\dfrac{P \qquad [P \to \bot]^1}{\bot}}{(P \to \bot) \to \bot} \to_E \atop \to_{I_1} \qquad\qquad \frac{\dfrac{\dfrac{[P]^2 \qquad [P \to \bot]^1}{\bot}}{(P \to \bot) \to \bot}}{P \to ((P \to \bot) \to \bot)} {\to_E \atop \to_{I_1} \atop \to_{I_2}}$$

In the leftmost derivation we have applied the $\to_E$-rule to the assumptions P and $P \to \bot$ and introduced $\bot$. Both P and $P \to \bot$ are open assumptions, and $\bot$ is the conclusion. The derivation expresses that if we assume P and $P \to \bot$, we get $\bot$. *In the middle derivation* we have applied the $\to_I$-rule to $P \to \bot$ and $\bot$. We have closed $P \to \bot$ and introduced $(P \to \bot) \to \bot$. The whole derivation expresses that if we assume P, we get $(P \to \bot) \to \bot$. The assumption P is still open. *In the rightmost derivation* we have applied the $\to_I$-rule to P and $(P \to \bot) \to \bot$. We have closed P and introduced $P \to ((P \to \bot) \to \bot)$. Now all the assumptions are closed. The whole derivation expresses that $P \to ((P \to \bot) \to \bot)$ holds no matter what we assume. ◆

Q **Derivation.** The following is a derivation of $(P \to (Q \to R)) \to ((P \land Q) \to R)$ in which all assumptions are closed:

$$\frac{\dfrac{\dfrac{[P \land Q]^1}{Q} \land_E \qquad \dfrac{\dfrac{[P \land Q]^1}{P} \land_E \qquad [P \to (Q \to R)]^2}{Q \to R} \to_E}{R} \to_E}{\dfrac{(P \land Q) \to R}{(P \to (Q \to R)) \to ((P \land Q) \to R)}} {\to_{I_1} \atop \to_{I_2}}$$

You should try to write down this derivation yourself, on paper, to understand how it is constructed step by step. *What does the conclusion of the derivation express? Is this a valid formula?* ◆

Derivations and Proofs

We see that all the derivations have a tree structure. It is also possible to write derivations in natural deduction as a list of formulas, but the structure appears more clearly with trees. We are now ready to define the set of derivations and the set of proofs, inductively. In practice, it is useful to think both "from the bottom up" and "from the top down" when we are trying to find proofs. We may well start with the premises and apply the rules in order to arrive at the desired conclusion. But we may also start with the conclusion and reason backward in order to find what the premises must be, by asking, "which rule was the most recently applied to reach this conclusion?"

> **Definition 24.1. Derivations and proofs**
>
> The set of **derivations** of natural deduction is inductively defined. The base set consists of the set of all propositional formulas, and there is one inductive step for each rule. Each rule tells how a new derivation can be constructed from one or two smaller derivations. A **proof** of a formula F is a derivation in which F is the conclusion and all the assumptions are closed. A formula is **provable** if there is a proof of it.

Q   **Proof.** The following shows how a *proof* of $P \wedge Q \rightarrow Q$ may be constructed step by step:

$$P \wedge Q \qquad\qquad \dfrac{P \wedge Q}{Q}\,{\wedge\text{E}} \qquad\qquad \dfrac{\dfrac{[P \wedge Q]^1}{Q}\,{\wedge\text{E}}}{P \wedge Q \rightarrow Q}\,{\rightarrow\text{I}_1}$$

We begin with $P \wedge Q$ and get Q by applying $\wedge\text{E}$. In the middle we have a derivation in which $P \wedge Q$ is an open assumption and Q is a conclusion. Then we can apply the $\rightarrow$I-rule to these formulas, introduce $(P \wedge Q) \rightarrow Q$, and *close* $(P \wedge Q)$. This is how we get the rightmost derivation, which is a *proof* because there are no open assumptions.   ◆

Q   **Proof.** The left-hand derivation below is not a proof; it is a derivation in which $P \rightarrow \bot$ is the conclusion and $P \rightarrow Q$ and $Q \rightarrow \bot$ are open assumptions. We have seen that $\neg F$ can be used as an abbreviation for $F \rightarrow \bot$, and the right-hand derivation is the same as the one on the left, but here we have used this abbreviation. In this derivation, we have continued with an extra $\rightarrow$-introduction and arrived at a derivation with $\neg Q \rightarrow \neg P$ as conclusion:

$$\dfrac{\dfrac{\dfrac{[P]^1 \quad P \rightarrow Q}{Q}\,{\rightarrow\text{E}} \quad Q \rightarrow \bot}{\bot}\,{\rightarrow\text{E}}}{P \rightarrow \bot}\,{\rightarrow\text{I}_1} \qquad\qquad \dfrac{\dfrac{\dfrac{\dfrac{[P]^1 \quad P \rightarrow Q}{Q}\,{\rightarrow\text{E}} \quad [\neg Q]^2}{\bot}\,{\rightarrow\text{E}}}{\neg P}\,{\rightarrow\text{I}_1}}{\neg Q \rightarrow \neg P}\,{\rightarrow\text{I}_2}$$

The right-hand derivation is still not a proof, because $P \rightarrow Q$ is an open assumption of the derivation. The derivation expresses that *if* $P \rightarrow Q$ is true, *then* $\neg Q \rightarrow \neg P$ must also be true.   ◆

Q **A special case.** We may freely apply the $\to$I-rule although no assumptions are closed. The following shows how a *proof* of $P \to (Q \to P)$ is constructed:

$$P \qquad\qquad \frac{P}{Q \to P}\,\to\text{I} \qquad\qquad \frac{\dfrac{[P]^1}{Q \to P}\,\to\text{I}}{P \to (Q \to P)}\,\to\text{I}_1$$

In the left derivation, we start with P. This is a derivation in which P is both an open assumption and the conclusion. By applying the $\to$I-rule once, we get $Q \to P$. No assumption Q is closed, but that is OK. Intuitively, this corresponds to the fact that if we know P, we also know $Q \to P$, no matter what Q is. In the right derivation we have closed the assumption P, giving us a *proof*.   ◆

Negation and RAA

There is an important difference between the following proofs:

$$\begin{array}{c} [F] \\ \vdots \\ \dfrac{\bot}{\neg F}\,\to\text{I} \end{array} \qquad\qquad\qquad \begin{array}{c} [\neg F] \\ \vdots \\ \dfrac{\bot}{F}\,\text{RAA} \end{array}$$

In the left-hand derivation, the assumption F leads to a contradiction, and therefore that F cannot be the case. In the right-hand derivation, the assumption $\neg F$ leads to a contradiction, but here *the conclusion is that F holds*, not only that $\neg F$ cannot be the case.

Q **Proof.** The following shows how a *proof* of $\neg\neg P \to P$ is constructed:

$$\frac{\neg P \qquad \neg\neg P}{\bot}\,\to\text{E} \qquad\quad \frac{\dfrac{[\neg P]^1 \qquad \neg\neg P}{\bot}\,\to\text{E}}{P}\,\text{RAA}_1 \qquad\quad \frac{\dfrac{\dfrac{[\neg P]^1 \qquad [\neg\neg P]^2}{\bot}\,\to\text{E}}{P}\,\text{RAA}_1}{\neg\neg P \to P}\,\to\text{I}_2$$

Without the RAA-rule it is impossible to prove $\neg\neg P \to P$. Feel free to try to prove $\neg\neg P \to P$ without the RAA-rule, but do not spend too much time trying.   ◆

Natural deduction without the RAA-rule is called **intuitionistic logic**. The example above shows that $\neg\neg P \to P$ is provable in classical logic using RAA, but it is not provable in intuitionistic logic. This logic is both historically important and useful in practice: a proof in intuitionistic logic is *constructive* in the sense that when we have an intuitionistic proof, we have access to more information than we have with a classical proof.

Digression

Natural deduction was developed in the 1930s. Before then, most logical systems were *axiomatic* in the sense that they consisted of many axioms and few inference rules. In 1934, two articles were published, independently of each other, by the Polish logician *Stanisław Jaśkowski* (1906–1965) and by the German mathematician and logician *Gerhard Gentzen* (1909–1945). These papers formed the starting point for natural deduction and what is today called *structural proof theory*, which is still actively researched. It was Gentzen who first used the term "natural deduction." He also introduced his **sequent calculus** in 1934, another logical calculus, which he used, among other things, as a tool for reasoning about natural deduction. The rules of sequent calculi typically look like this:

$$\frac{\Gamma, F \vdash G}{\Gamma \vdash F \to G} \qquad\qquad \frac{\Gamma \vdash F \qquad G, \Gamma \vdash H}{\Gamma, F \to G \vdash H}$$

The left-hand rule can be viewed as another formulation of the $\to$I-rule in natural deduction: "if from the assumptions in Γ and F you have a proof of G, then you have from the assumptions in Γ a proof of $F \to G$." Intuitively, this corresponds to closing the assumption F in the derivation of $F \to G$. The right-hand rule is a little more complex, but it can be read in a similar way.

The Rules for Disjunction

We can express everything we need with the connectives $\to$, $\wedge$, and $\bot$. In fact, we need only $\to$ and $\bot$, because $\wedge$ can be defined in terms of them. It is nevertheless interesting to look at the disjunction and $\vee$-formulas. It is possible to look at $F \vee G$ as an abbreviation of $\neg(\neg F \wedge \neg G)$, which is perfectly fine, but here we will look at how we can define rules directly in natural deduction for $\vee$-formulas. Here are the rules for $\vee$-formulas:

$$\frac{F}{F \vee G} \vee \text{I} \qquad\qquad \frac{G}{F \vee G} \vee \text{I} \qquad\qquad \frac{F \vee G \qquad \begin{array}{c}[F]\\ \vdots\\ H\end{array} \qquad \begin{array}{c}[G]\\ \vdots\\ H\end{array}}{H} \vee \text{E}$$

We have two introduction rules. The first one says that if we have a proof of F, we also have a proof of $F \vee G$, and the other says the same thing for G. The elimination rule intuitively states that if we know $F \vee G$, and from both F and G we can conclude H, then we can conclude H and close the assumptions F and G. This is a formalization of *proof by cases*, which we met in Chapter 5 (*page 55*).

Q   Proofs. The following are proofs of $P \to (P \vee Q)$ and $P \vee Q \to Q \vee P$, respectively:

$$\cfrac{\cfrac{[P]^1}{P \vee Q} \vee_I}{P \to (P \vee Q)} \to_{I_1} \qquad\qquad \cfrac{\cfrac{[P \vee Q]^2 \quad \cfrac{[P]^1}{Q \vee P}\vee_I \quad \cfrac{[Q]^1}{Q \vee P}\vee_I}{Q \vee P}\vee_{E_1}}{P \vee Q \to Q \vee P} \to_{I_2}$$

$\blacklozenge$

Soundness, Completeness, and Consistency

A *proof* of a formula F is defined as a derivation in which F is the conclusion and all the assumptions are closed, and a formula is defined as *provable* if there is a proof of it. The set of all provable formulas of natural deduction is thus also defined, and it turns out that this set is exactly the same as the set of *valid* formulas! This means that we have managed to give a *syntactic* characterization of the valid formulas:

- If a formula is provable, is it valid (**soundness**).
- If a formula is valid, it is provable (**completeness**).

The properties *soundness* and *completeness* are absolutely central to logic, and when logical calculi are defined, these are two of the most fundamental properties we almost always want to prove.

Definition 24.2. Soundness and completeness

A calculus is **sound** if each provable formula in the calculus is valid. A calculus is **complete** if each valid formula is provable in the calculus.

Soundness and completeness are dual concepts, and they build an important bridge between syntax and semantics. Provability is a syntactic property, and validity is a semantic property. Soundness means that we cannot prove anything more than the valid formulas – that is, the calculus is *correct* – and completeness means that we can prove *all* valid formulas – that is, the calculus is *complete*.

Natural deduction is both sound and complete, but it is easy to modify it so that these properties are lost: if we remove a rule, for example RAA, we lose completeness, and if we add an extra rule, we quickly lose soundness.

In general, it is easier to prove soundness than completeness: To prove that a calculus is sound, we assume that a formula is provable and show that it must be valid. What makes soundness easier to prove is that we work on a specific object, a proof of the formula, and reason about this object. We can also prove soundness contrapositively by assuming that a formula is *not* valid and showing that it is *not* provable. That a formula is not valid means that it has a countermodel, but then we also have a specific object to reason about, in this case a model or valuation. On the other hand, to prove that a calculus is complete, we start with a universal assumption, either that a formula is valid or that a formula is *not* provable. It is generally more difficult than proving soundness, for we must here *construct* a proof or a countermodel. We will not prove soundness and completeness here, but both proofs are done using structural induction, one on the set of derivations and the other on the set of formulas. We can summarize the relationship between validity and provability in this way:

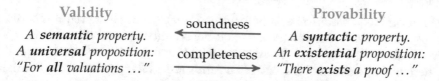

Validity		Provability
*A **semantic** property.*	soundness ←	*A **syntactic** property.*
*A **universal** proposition:*	completeness →	*An **existential** proposition:*
*"For **all** valuations …"*		*"There **exists** a proof …"*

Q **Soundness and completeness of procedure.** The concepts of soundness and completeness can also be used in other contexts. Suppose that we are writing a program for printing out prime numbers. Then the soundness and completeness of the procedure are the following:

– **Soundness** of the procedure: *Everything that is printed is a prime number.*
– **Completeness** of the procedure: *All prime numbers are printed.*

A procedure may well be sound without being complete, and then it does too little, for example if it prints only the numbers 2, 5, 7, 11, 17, 19, …. Here, the prime numbers 3 and 13 are missing, but all the numbers in the list are actually prime numbers. A procedure can also be complete without being sound, and then it does too much, for example if it prints the numbers 2, 3, 5, 7, 9, 11, 13, 15, 17, 19, 21, …. Here are all the prime numbers, but also 9, 15, and 21, which are not prime numbers. As a rule, we want both soundness and completeness, and that neither too much nor too little is printed: 2, 3, 5, 7, 11, 13, 17, 19, ….   ◆

All the concepts we are talking about here apply to both propositional logic and first-order logic. In Chapter 16 (*page 186*) we defined the concepts of *theory, axiom,* and *theorem* for first-order logic, and using *derivations* and *proofs* in natural deduction, we have now defined the syntactic counterparts of *logical consequence* and *logical equivalence*. We can make the connection very clear by defining *derivability* of formulas:

Definition 24.3. Derivability

Let M be a set of formulas. We say that a formula F is **derivable**, and that it can be **derived** from M, if there is a derivation of F in which all the open assumptions are formulas of M. In that case, we write M ⊢ F, and we write ⊢ F if there is a proof of F.

We can now express soundness and completeness in the following way:

Soundness: $M \vdash F \quad \Rightarrow \quad M \models F$
Completeness: $M \models F \quad \Rightarrow \quad M \vdash F$

Here, soundness and completeness are expressed by means of symbols, but it is important to be aware that the symbols are only an abbreviation of something we can say with very ordinary words. This is exactly what we mean by saying that the symbols belong to the *metalanguage* and not the *object language*.

We have defined a *theory* as a set of formulas, without any extra requirements, but it is also common to define a theory to be closed under provability, that is, if $T \vdash F$, then F is also in the theory T. An essential property of a theory, when derivability is defined, is that we cannot derive both a formula and its negation from the theory:

Definition 24.4. Consistent theory

A set of formulas is **consistent** if is impossible to derive both a formula and its negation from it.

Using this concept, we can now formulate soundness and completeness as follows:

Soundness: Every satisfiable set is consistent.
Completeness: Every consistent set is satisfiable.

To show that a theory is consistent, it is sufficient to find a model that satisfies the theory. If the underlying calculus is sound, it follows that the theory is consistent. There is much more to be said about logic, but here we must draw the line.

Digression

The Austrian logician and mathematician *Kurt Gödel* (1906–1978) was one of the most significant logicians ever, and his work has had tremendous influence on logic, mathematics, and philosophy. It was he who first, in 1929, proved the completeness of a calculus for first-order logic, but he is perhaps best known for his two incompleteness theorems from 1931. The first is that for every sufficiently strong consistent axiom system for number theory, there will be statements that are true but not provable from the axioms. To prove this, Gödel developed a method for formalizing statements of the form "this sentence is not provable." The second incompleteness theorem intuitively states that every sufficiently strong consistent axiom system for number theory is unable to prove its own consistency.

Exercises

24.1 Explain in your own words what natural deduction is and what it can be used for.

24.2 Find at least four different ways in which to *read* the following rule:

$$\frac{A \qquad B}{A \wedge B} \ \wedge\text{I}$$

Hint: You can read it syntactically or semantically, and you can read it from above or below.

24.3 There are no references to *truth* or *falsity* in the rules of natural deduction. The rules are *syntactic* and tell us only how to construct *derivations* and *proofs*. Why is that so? What are the semantic counterparts of *derivability* and *provability*?

24.4 The rules of natural deduction describe ways to *reason* and *prove*. What connections can you find between the rules of natural deduction and the proof methods in Chapter 5?

24.5 For each of the following formulas, find a *proof* in natural deduction for the formula. No assumptions should be open in the final proof.

(a) $(A \wedge B) \to A$

(b) $(A \wedge B) \to (B \wedge A)$

(c) $(A \wedge (B \wedge C)) \to (A \wedge C)$

(d) $A \to (B \to (A \wedge B))$

(e) $(P \wedge Q) \to (P \vee R)$

(f) $(P \to Q) \wedge (R \to Q) \to (P \wedge R) \to Q$

(g) $(P \to Q) \wedge (Q \to R) \to (P \to R)$

(h) $(P \wedge Q \to R) \to P \to Q \to R$

(i) $P \to P$

(j) $P \to \neg\neg P$

(k) $(P \to Q \to R) \to (P \wedge Q \to R)$

(l) $(P \to P \to Q) \to (P \to Q)$

(m) $\neg\neg P \to P$

(n) $(A \to B) \to (\neg B \to \neg A)$

24.6 Find proofs of the following formulas. These are a little more difficult than those in the previous exercise.

(a) $\neg P \to \neg(P \wedge Q)$

(b) $\neg P \to \neg Q \to \neg(P \vee Q)$

(c) $(P \wedge \neg P) \to Q$

(d) $(P \wedge (Q \vee R)) \to ((P \wedge Q) \vee (P \wedge R))$

(e) $((P \wedge Q) \vee (P \wedge R)) \to (P \wedge (Q \vee R))$

(f) $((P \to Q) \to P) \to P$

24.7 The goal of this exercise is to provide a proof of the formula $(P \vee \neg P)$. To make it a little easier, there are two formulas that it pays to try first. Then these proofs can be combined into a proof of $(P \vee \neg P)$.

(a) Find a derivation of $\neg P$ in which $\neg(P \vee \neg P)$ is an open assumption.

(b) Find a derivation of $\neg\neg P$ in which $\neg(P \vee \neg P)$ is an open assumption.

(c) Combine these two derivations to obtain a proof of $(P \vee \neg P)$.

24.8 Find proofs of the following formulas. These require a little more pondering. Hint: It often becomes easier if you use previous proofs, for example the proof of $(P \vee \neg P)$.

(a) $\neg(P \wedge Q) \to (\neg P \vee \neg Q)$

(b) $\neg(P \vee Q) \to (\neg P \wedge \neg Q)$

(c) $(P \to Q) \vee (Q \to P)$

(d) $(P \to Q) \to (\neg P \vee Q)$

24.9 Now we will modify the calculus by adding a new rule. Let us define a rule that removes double negations, as in the top right-hand figure below.

We call this the $\neg\neg$-rule. We will now argue that we cannot prove anything *more* by adding this rule. We do this by translating proofs with the RAA-rule to proofs with the $\neg\neg$-rule and vice versa:

$$\frac{\neg\neg F}{F} \; \neg\neg$$

(a) First, suppose you have a proof in which the RAA-rule is applied only at the end, that is, at the very bottom, as in the middle figure to the right. Construct a proof for F in which the RAA-rule does not occur.

$$[\neg F]^1$$
$$\vdots$$
$$\frac{\bot}{F} \; \text{RAA}_1$$

(b) Suppose you have a proof in which the $\neg\neg$-rule is applied only at the end, as in the lower figure to the right. Construct a proof of F in which the $\neg\neg$-rule does not occur.

$$\vdots$$
$$\frac{\neg\neg F}{F} \; \neg\neg$$

24.10 For each of the following formulas, provide a proof in intuitionistic logic of the formula or argue why such a proof cannot exist.

(a) $P \to (Q \to P)$

(b) $P \to (Q \to (Q \wedge P))$

(c) $\neg\neg\neg P \to \neg P$

(d) $((P \to Q) \to P) \to P$

24.11 Find a theory $\{F, G, H\}$ that is not consistent but such that every subset of two or fewer elements is consistent.

24.12 Use your imagination to find an example of a procedure or method that is:

(a) *sound*, but not *complete*

(b) *complete*, but not *sound*

(c) neither *sound* nor *complete*

24.13 Prove that "every satisfiable set is consistent" and "every provable formula is valid" are equivalent statements. Explain in your own words why this makes sense.

24.14 Prove by structural induction on the set of derivations that natural deduction is a *sound* calculus, that is, that if $M \vdash F$, then $M \models F$.

The Road Ahead

Thanks for reading. I hope you have learned a lot, and, not least, had a fun and challenging experience along the way. If anyone had asked me what I most wanted you to be left with after reading this book, it's a *feeling for* and an *experience with* the following, which make up some of the *essence* of mathematics, as I see it:

- The distinction between syntax and semantics.
- The interaction between the simple and the complex.
- The freedom that lies in allowing us to assume anything.
- The art of thinking abstractly and mathematically.

Included in these four points is something we do all the time in math: we observe and abstract over what we observe, we find patterns, we represent what we find in a suitable language, and we apply algorithms and methods to the representations and thereby gain greater insight and understanding.

Some book tips. Finally, I want to recommend some books. There are many good books on these topics, and it is difficult to choose. The following books are some of my favorites, and much of this book draws its inspiration from them. *Good reading and until next time!*

The Classics

George Polya. *How to Solve It*. Princeton University Press, 1945. This is an incredibly good book that has sold over one million copies and been translated into over 17 languages. Actually, you should put away the book you have in front of you now and read *How to Solve It*. This is a book on how to solve mathematical problems. In Chapter 0 (*page 4*), we described some of the tips and rules of thumb from this book.

Donald E. Knuth. *The Art of Computer Programming*. Addison-Wesley, 1997. This is a fundamental work in computer science that covers most aspects of algorithms and programming. Knuth began this series in 1962, and Volume 5 is scheduled for 2025. This is a book series that I cannot recommend highly enough.

Ronald L. Graham, Donald E. Knuth, and Oren Patashnik. *Concrete Mathematics: A Foundation for Computer Science*. Addison-Wesley, 1994. This is a gem of a book, but it requires some mathematical maturity. It is especially interesting if you want to learn basic mathematics in order to analyze and reason about the complexity of algorithms.

Harold Abelson and Gerald Jay Sussman. *Structure and Interpretation of Computer Programs*. MIT Press/McGraw-Hill, 1996. This is probably one of the very best introduc-

© Springer Nature Switzerland AG 2021
R. Antonsen, *Logical Methods*, https://doi.org/10.1007/978-3-030-63777-4_26

tory books on programming available, and it is used worldwide. The principles and methods you will learn in this book will benefit you greatly, whether you go into mathematics or computer science. The book is freely available online.

Douglas R. Hofstadter. *Gödel, Escher, Bach*. Basic Books, 1999. This playful and strange book was described as "a metaphorical fugue on minds and machines in the spirit of Lewis Carroll" when it came out in 1979, and it explores the relationships between the logician *Kurt Gödel*, the artist *M. C. Escher*, and the composer *Johann Sebastian Bach*.

John Horton Conway and Richard K. Guy. *The Book of Numbers*. Springer, 1996. This is a wonderful book on *numbers*, and it has more figures than words. The authors manage in an original and understandable way to convey deep truths about numbers and mathematics.

Introductory Books on Mathematical Thinking

Keith Devlin. *Introduction to Mathematical Thinking*. Theoklesia, LLC, 2012. A short, good, and easy-to-read book written by the British-American mathematician and science communicator *Keith Devlin* (1947–) to facilitate the transition to university mathematics. Devlin is a very good communicator of mathematics.

Carol Schumacher. *Chapter Zero: Fundamental Notions of Abstract Mathematics*. Addison-Wesley, 2001. This is a gentle and careful introduction to mathematical thinking and mathematical proofs. The book focuses on mathematical structures and proofs.

Richard Hammack. *Book of Proof*. Creative Commons, 2013. This is a book that is both freely available online and is a great supplement to this book, precisely because it is about the same thing: how to do mathematical proofs on your own. My illustration of dominoes falling was inspired by a similar image in this book. Download the book and have a look.

Introductory Books on Logic

Richard Jeffrey and John Burgess. *Formal Logic: Its Scope and Limits*. Hackett, 2006. This is a good, concise introduction to logic that is based on natural deduction. The book may be difficult for a beginner, but you are no longer one, so it will go well!

David Barker-Plummer, Jon Barwise, and John Etchemendy. *Language, Proof, and Logic*. CSLI, 2011. This is a thorough textbook on logic, where the focus is on proof, methods of proof, and the relationship between language and logic. The book is complemented by its own software.

Dirk van Dalen. *Logic and Structure*. Springer, 2004. This is an introduction to mathematical logic based on natural deduction. The book is somewhat more mathematically demanding than the others on the list, but if you are comfortable with structural induction and natural deduction, this is the book for you.

Raymond M. Smullyan. *Logical Labyrinths*. AK Peters, 2009. This is a fairly new introductory book in logic, which merges entertaining logic puzzles with theory. Highly recommended.

Patrick Blackburn, Maarten de Rijke, and Yde Venema. *Modal Logic*. Cambridge University Press, 2001. This is my favorite logic book. It is an introduction to *modal logic*, the logic for talking about relational structures. It is well written and contains a lot of beautiful mathematics, but it requires some mathematical maturity.

Introductory Books on Discrete Mathematics

John T. O'Donnell, Cordelia V. Hall, and Rex L. Page. *Discrete Mathematics Using a Computer*. Springer, 2006. This is a great book if you want to learn math, logic, *and* programming at the same time. It is an introductory book that combines programming in *Haskell* with mathematical reasoning and proof writing.

Kees Doets and Jan van Eijck. *The Haskell Road to Logic, Maths and Programming*. King's College Publications, 2004. This book is thematically quite similar to the book above; it is also an introductory book for logic and mathematics via the Haskell programming language. But it is perhaps somewhat more demanding.

James Hein. *Discrete Structures, Logic, and Computability*. Jones & Bartlett, 2010. This is a very comprehensive book that aims to form a thorough foundation for computer science students, and it does. It works well as an encyclopedia and has been used in similar courses at the University of Oslo in the past.

Ron Haggarty. *Discrete Mathematics for Computing*. Addison-Wesley, 2002. Peter Grossman. *Discrete Mathematics for Computing*. Palgrave Macmillan, 2009. Kenneth Rosen. *Discrete Mathematics and Its Applications*. McGraw-Hill, 2012. These are three good introductory books on discrete mathematics. The first two are written for computer science students and are quite short; the third is more comprehensive.

Popular Science, Recreational Mathematics, and Other Books

Martin Gardner. *The Colossal Book of Mathematics*. Norton, 2001. The American mathematician and science writer *Martin Gardner* (1914–2012) is a legend in recreational mathematics, and he wrote over 100 books over a period of 80 years. This book is a collection of articles and papers that Gardner wrote for *Scientific American* from 1956 to 1981.

Ian Stewart. *The Great Mathematical Problems*. Profile, 2013. The British mathematician *Ian Stewart* (1945–) has been writing popular science books on math for over 30 years, and you can pick up almost any of them and have a great experience. This book is one of the latest and deals with the great unresolved problems of mathematics.

John Stillwell. *Yearning for the Impossible: The Surprising Truths of Mathematics*. AK Peters, 2006. This is a well-written and inspiring book about how we strive for the "impossible" in mathematics. The book succeeds in explaining more advanced mathematics than many other popular science books.

Paul Lockhart. *Measurement*. Harvard University Press, 2012. In 2002, the American mathematics teacher Paul Lockhart wrote a pamphlet entitled "A Mathematician's Lament," in which he compares mathematics education in school with music education without music.

In *this* book he manages to convey both the beauty of mathematical proofs and the pleasure of exploring mathematics on your own.

Peter Winkler. *Mathematical Puzzles: A Connoisseur's Collection*. AK Peters, 2004. This book is a little dangerous, because if you open it and read one of the puzzles, you run the risk of being stuck for a long time. Not necessarily because the puzzles are difficult, but because they are good. If you like mathematical puzzles of high *quality*, then this is the book for you.

A. Doxiadis, C.H. Papadimitriou, A. Papadatos, and A. Di Donna. *Logicomix: An Epic Search for Truth*. Bloomsbury, 2009. This is a *comic book* about logic in which we get an insight into the life of the British mathematician and philosopher *Bertrand Russell* (1872–1970) and his attempt to understand the world.

Robin J. Wilson. *Introduction to Graph Theory*. Prentice Hall, 2010. This is a great and classic introduction to graph theory that is highly recommended. The book was first published in 1972, but there have been many updated editions since then. If you found graph theory interesting, take a look at this one.

Raymond M. Smullyan. *The Lady or the Tiger?* Alfred A. Knopf, 1982. The American logician *Raymond Smullyan* (1919–2017) is perhaps best known for his puzzle books, and this is one of them. Here you can test yourself on hundreds of simple puzzles.

Raymond M. Smullyan. *Forever Undecided: A Puzzle Guide to Gödel*. Oxford University Press, 1987. This book explains Gödel's fundamental result on incompleteness in an original and understandable way. Instead of talking directly about *logics* and *provability*, Smulluyan creates a universe of people who *reason* and have *beliefs* about others and themselves. A lovely book!

P.J. Davis, R. Hersh, and E.A. Marchisotto. *The Mathematical Experience*. Birkhäuser, 2012. This book tries to shed some light on *what mathematics really is* in both a philosophical, but also historical, perspective. The book beautifully describes what the humane and essential is in being mathematical.

Index

© Springer Nature Switzerland AG 2021
R. Antonsen, *Logical Methods*, https://doi.org/10.1007/978-3-030-63777-4

Symbols

© Springer Nature Switzerland AG 2021
R. Antonsen, *Logical Methods*, https://doi.org/10.1007/978-3-030-63777-4

Printed in the United States
by Baker & Taylor Publisher Services